Peter Ax
Multicellular Animals
A new Approach to the Phylogenetic Order in Nature

Volume I

AKADEMIE DER WISSENSCHAFTEN UND DER LITERATUR · MAINZ

Springer

Berlin
Heidelberg
New York
Barcelona
Budapest
Hong Kong
London
Milan
Paris
Santa Clara
Singapur
Tokyo

Peter Ax

Multicellular Animals

A new Approach
to the Phylogenetic Order in Nature

Volume I

Springer

Original edition: Peter Ax. Das System der Metazoa I. Ein Lehrbuch der phylogenetischen Systematik.
Gustav Fischer Verlag, Stuttgart, Jena, New York.
Akademie der Wissenschaften und der Literatur, Mainz 1995

Prof. Dr. Peter Ax, Universität Göttingen, II. Zoologisches Institut und Museum
Berliner Straße 28, D-37073 Göttingen

Translated by Michael Power, Göttingen

Book cover:
Gnathostomulida – an example of the ground pattern of descent communities composed of autapomorphies and plesiomorphies.
Center: Anterior end of *Gnathostomula paradoxa* with cuticular basal plate and jaw apparatus in a muscular pharynx (autapomorphy).
Top: Skin cells of *Gnathostomaria lutheri* with one cilium per cell (plesiomorphy).
Original state of the epidermis in the Bilateria.

Sponsored by Bundesministerium für Bildung, Wissenschaft, Forschung und Technologie, Bonn,
and by Niedersächsisches Ministerium für Wissenschaft und Kultur, Hannover.

ISBN 3-540-60803-6 Springer-Verlag Berlin Heidelberg New York

Cover Design: Erich Kirchner, Heidelberg
Typesetting and Printing: Röhm GmbH, Sindelfingen
SPIN 10529967 31/3137 – 5 4 3 2 1 0 – printed on acid free paper

Contents

Introduction

No one can ever have secure knowledge about the gods and creatures, and should anyone hit by chance upon the right thing, he will not know it for sure; that is why everything that we believe to be true is "opinion".

XENOPHANES around 500 B.C.
(According to RÖD 1988, p.85)

The goal of phylogenetic systematics (cladistics) is to discover the kinship relations between all organisms on earth and to translate the order we perceive in Nature into an equivalent man-made system.

Although the goal is easily formulated, the path is thorny, and the results achieved continue to be imperfect. This is the fate of any science that bases its propositions on the interpretation of historical evidence.

The diversity found in the millions of species originated as a result of the continuous splitting of biopopulations through time. Combined with this was the emergence of hierarchically linked descent communities of species. We call the process of origin of descent communities phylogenesis. We do not know, however, the exact course of phylogenesis – we can only formulate hypotheses.

The historical evidence at hand consists of the feature patterns of extant species and of extinct species with their combination of original and derived traits which are the result of evolution.

The theory of phylogenetic systematics (HENNIG 1950, 1966) and the methods of evaluating feature patterns are discussed in German textbook literature (AX 1984, 1988; SUDHAUS and REHFELD 1992). The subject matter of the book at hand is the practice of systematization, which in this case applies to the Metazoa, a huge descent community of heterotrophic multicellular organisms, often referred to simply as animals. The unicellular heterotrophs grouped under the name Protozoa, however, are not the equivalent of a descent community in Nature.

It should be possible to read THE SYSTEM OF THE METAZOA independently of other works. For this reason, the introductory chapters deal with the composition of life in Nature and with the path leading to the phylogenetic system as a construct of man. Some questionable, broad classifications of organisms are discussed in a critical light.

In the first chapter we will consider the ontology compiled by BUNGE (1977, 1979). In his opinion, only individuals and populations can be interpreted as things or material systems. Species and descent communities, on the other hand, are natural kinds with characteristics of classes.

This book seeks a new path in the field of teaching. It is designed as an alternative to traditional textbooks, which still include typological divisions and use categories originating from Linné.

THE SYSTEM OF THE METAZOA is neither more nor less than a framework of well-founded and compatible hypotheses which may at any time be confirmed or rejected by empirical testing. It does not deal with an unchangeable, monolithic structure that we learn once and can keep as a convenient pigeonhole for the rest of our lives. New findings will be continuously incorporated into the system, and these will necessitate continuous revisions and changes.

The phylogenetic system of any group of organisms can be depicted in two different ways, both giving identical information – as a diagram of phylogenetic relationships, and in the form of a hierarchical tabulation that will be discussed in detail later. There is persistent opposition to this view. The methods of phylogenetic systematics might be useful to uncover kinship relationships in Nature and to depict them in relationship diagrams or cladograms; but this is claimed to be as far as phylogenetic systematics can go. The written system of organisms, on the other hand, should take the results of all biological research into consideration, for example, the extent of evolutionary change with regard to the delimitation of taxa. Apart from the fact that no objective criteria exist for this, the respective (selected) measure would have to be indicated in each case to call up the stored information intersubjectively – an utterly hopeless endeavor.

Should, contrary to current belief, further objective criteria for the systematization of living creatures be discovered (i.e., criteria not based on descent), the resulting systems would have to be written independently of the phylogenetic system.

Ms. Renate Grüneberg intuitively and circumspectively drew the kinship diagrams, labeled the pictures and managed the computer. Talent and skill guided the hand of Ms. Anne Theenhaus, who drew the illustrations from originals and copies taken from literature. Bernd Baumgart redrew the pictures of sponges and several other figures with reliable precision. My sincere thanks to you all.

I would also like to thank once again the Academy of Sciences and Literature in Mainz for their kind cooperation.

I am very greatful to the Springer publishing house for taking charge of the English edition.

Organisms of the Real World and Their Arrangement in a System Created by Man[1]

Hypothetical Realism

There is a real world which exists independently of our intellectual capacity to understand it. This is the central philosophy of realism.

If this is true, what, then, is the relationship between the real world and our intellectual capacity? The human cognitive apparatus hardly originated arbitrarily. It is more realistic to assume that during the course of evolution our perception adapted itself to the real world, and forms at least a partially correct image of it. From this point of view, it appears legitimate to assume that what we perceive more or less corresponds to reality. We can, however, never have certain knowledge about reality, and statements about it will always remain hypotheses, confined to the realms of "hypothetical realism" (VOLLMER 1975).

Levels of Observation

Let us consider Nature on the basis of these assumptions. What can we know about the composition of organisms in the real world at a certain place and a certain time? Take the example of jellyfish. Various large specimens can be seen on the coast of the North Sea – the common jellyfish, the red jelly, the compass jellyfish, and the cauliflower jellyfish (Fig. 1). They are often carried towards shore by currents and run the risk of being stranded on the beach at low tide.

We can observe these animals on three levels: we can study single individuals, trace populations comprising very similar individuals, or, finally, we can look at population groups that demonstrate certain congruences.

The Individual

The starting point of all observation is the individual, single organism capable of independent existence.

Let us take a closer look at one representative individual of each of the four kinds of jellyfish in question. The common jellyfish has a flat bell; short tentacles hang from the rim of the smooth umbrella; the gonads appear in the bell in four circles. The red jelly does not have tentacles at the lobular umbrella rim, but it has very long thread-like, stinging tentacles on the underside of the bell. The rim of the flat bell of the compass jellyfish has thick, long tentacles, and brown pigment

---

[1] I thank Dr. M. Mahner (Montreal) for the manuscript MAHNER and BUNGE, PHILOSOPHY of BIOLOGY and for suggestions on how to improve the chapter. Any remaining unclarities are my fault.

stripes run along the surface. The hemispherical cauliflower jellyfish has no tentacles at all. This is linked to its peculiar method of ingestion. A large oral opening such as is seen in the other jellyfish is missing, instead its oral arms are partially grown together. A dense system of ducts transports plankton from the shoulder and mouth frill areas to the stomach.

Population

Each individual is a unique being, occurring only once in Nature. Notwithstanding this, the four jellyfish in question are not unique phenomena in the North Sea, but can appear synchronically and syntopically in hundreds. Although a truism, these simple facts need to be stated if we are to approach the level of populations, which are groups of individuals bearing a great number of commonalities.

The individuals in each of the four jellyfish populations distinguish themselves by a regularly correlated combination of certain characteristics, some of which have already been mentioned. The phenotypic similarities between the individuals of a population are an expression of a correspondingly similar genotype; and this is the prerequisite for a new combination of genetic information arising from the individuals of a population in the course of sexual reproduction.

We can observe this process by placing male and female common jellyfish together in an aquarium. Sperm are freely released into the water. Swimming larvae develop from the fertilized eggs in incubation sacks in the mouth arms of the female. These planulae attach themselves to the substrate and grow into small polyps. The lifecycle is completed at this stage with the famous switch from polyp to medusa. Small disks are vegetatively produced at the terminal end of the polyp, releasing young medusae called ephyra larvae, which have eight large arms (Fig. 38).

Groups of Populations with Certain Common Features

On the third level, we compare the different jellyfish populations. Are there perhaps features which only they have in common – characteristics which are not present in other animals? As far as our example is concerned, this holds true for the structure of the polyp – four endodermal septa projecting into the intestinal cavity which are equipped with ectodermal muscle funnels (Fig. 37), and for the germination of medusae in the form of ephyrs described above.

Of course, these characteristics are found not only in the four jellyfish we chose from the North Sea, but also in other jellyfish from different oceans. What is important, however, is that not all jellyfish around the world have polyps with four septa, and ephyrs, but only a certain limited number. An explanation for this will be offered later.

For the sake of orderliness, it remains to be said that our observations of individuals, as well as our comparisons of characteristics within a population and between populations are possible on every level imaginable. This can begin with the entire form, and stretch from macroscopy and microscopy to the ultrastructure of cells and their parts. This may include elements of behavior or development, the method of nourishment, the molecular patterns of enzymes, or even the DNA configuration of the genetic information.

Thing and Class

Logic distinguishes between things (= individuals) as concrete systems of Nature composed of parts (components), and classes as constructs of man made up of members (elements).

Things are material objects which exist in the real world. **Classes**, however, are conceptual

Fig. 1. Four individuals of jellyfish populations from the North Sea. A. Common jellyfish (*Aurelia aurita*). B. Red jelly (*Cyanea capillata*). C. Compass jellyfish (*Chrysaora hysoscella*). D. Cauliflower jellyfish (*Rhizostoma octopus*). A and B swimming individuals from the shore areas; C and D organisms which were washed ashore.

13

objects without a corresponding reality; they are thought up, created by man to group together certain things.

Individuals and Populations
as Things

According to BUNGE's ontology (1977, 1979; MAHNER 1993, 1994; MAHNER and BUNGE, ms), in the biological world only the individual and the population consisting of individuals living together in the same place and time have the status of material objects with real existence (Fig. 2).
In the following passage, the term population refers to a reproductive community (biopopulation) and not to biocenoses or to social systems of organisms.

This statement requires no further clarification as far as the **individual** is concerned. Biological individuals are composite wholes. Organs, tissues, and cells build parts of a multicellular organism. Certain cell components form the material system of a unicellular organism.

In the case of **biopopulations**, it is necessary to look further back. In material systems, binding relationships exist between the parts. Parts of the system have an effect on other parts or are influenced by them, as is the case between the cells of a multicellular individual. The contact between individuals as sexual partners is, for example, a comparable relationship in populations. Material systems also have characteristics which are not exhibited by their individual parts. In biopopulations these are the characteristics of the common gene pool, which is the total of the genetic information of the population at a particular point in time. Mutations in the genetic information of single individuals can dominate in subsequent generations and thus, over time, entirely change the appearance of the population. Evolution takes place in biopopulations.

Species and Descent Communities
as Natural Kinds (Classes)

Species
The notion of species has – with good reason – not yet appeared in this argument. Biological species do not, according to BUNGE, have any ontological status; they are not things, but classes. Before this position can be explained in greater detail, let us first shift the focus of our discussion from populations to species.
Our four different populations of jellyfish are not only found in the North Sea. Comparable populations exist in other oceans, and the individuals of which they are composed also have ancestors and offspring. We are dealing here with the expansion of the nondimensional biopopulation in space and time. Species encompass – in space – all populations with extensively matching individuals, and they include – in time – all populations whose individuals succeed each other by means of reproduction. Species have a beginning and an end (Fig. 3). They usually, if not exclusively, begin with the splitting of populations of other species. Species become extinct through splitting processes within their own populations or when their individuals die out (WILLMANN 1985). At this point we come back to ontology. The interpretation of species as individuals in the sense of things (material systems) was promoted in particular by GHISELIN (1974a) and HULL (1976), and

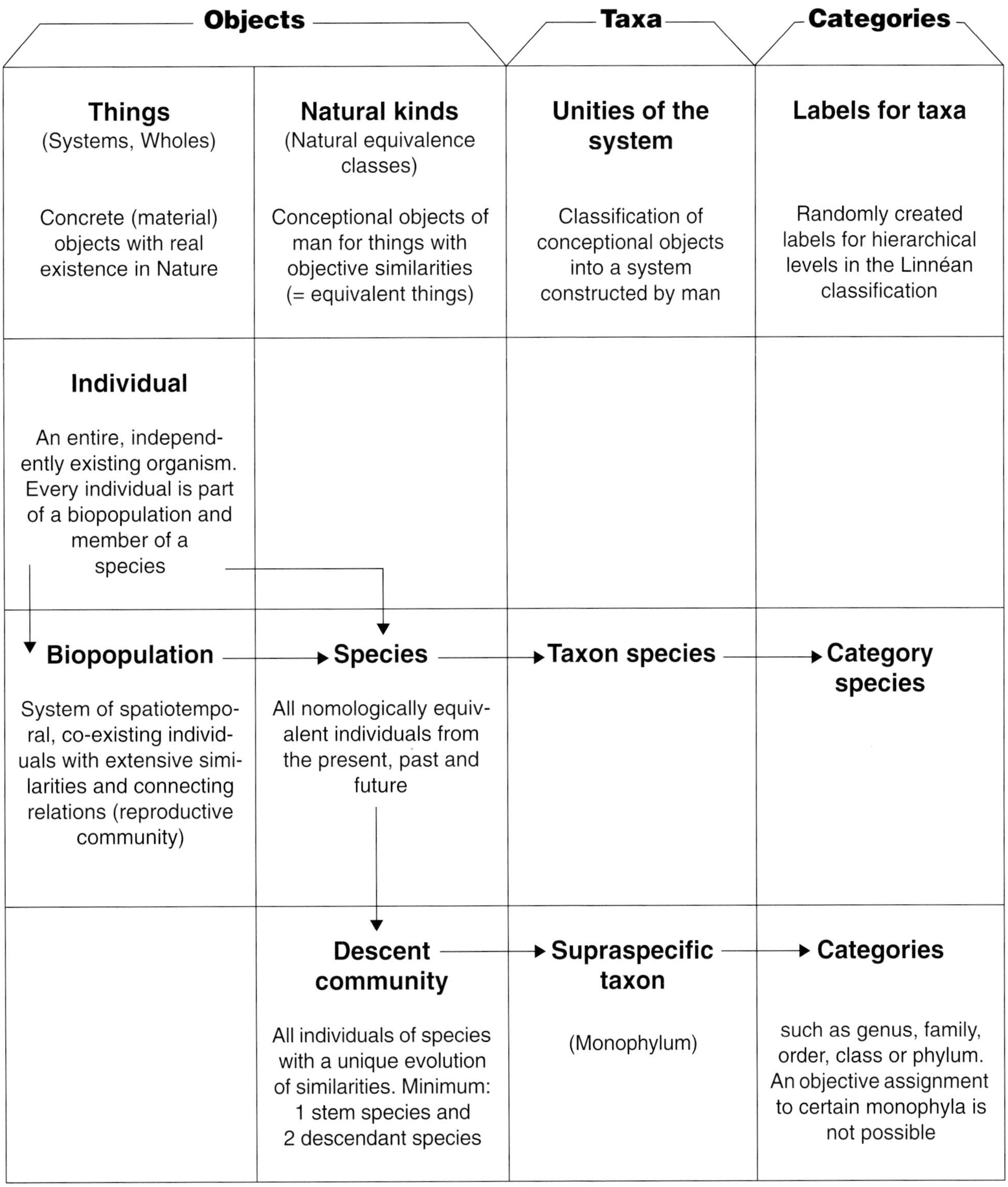

Objects		Taxa	Categories
Things (Systems, Wholes) Concrete (material) objects with real existence in Nature	**Natural kinds** (Natural equivalence classes) Conceptional objects of man for things with objective similarities (= equivalent things)	**Unities of the system** Classification of conceptional objects into a system constructed by man	**Labels for taxa** Randomly created labels for hierarchical levels in the Linnéan classification
Individual An entire, independently existing organism. Every individual is part of a biopopulation and member of a species			
Biopopulation System of spatiotemporal, co-existing individuals with extensive similarities and connecting relations (reproductive community)	**Species** All nomologically equivalent individuals from the present, past and future	**Taxon species**	**Category species**
	Descent community All individuals of species with a unique evolution of similarities. Minimum: 1 stem species and 2 descendant species	**Supraspecific taxon** (Monophylum)	**Categories** such as genus, family, order, class or phylum. An objective assignment to certain monophyla is not possible

Fig. 2. Relations between concrete objects of Nature, human conceptional objects, taxa of the phylogenetic system, and categories of the Linnéan hierarchy. Details are found in the text

15

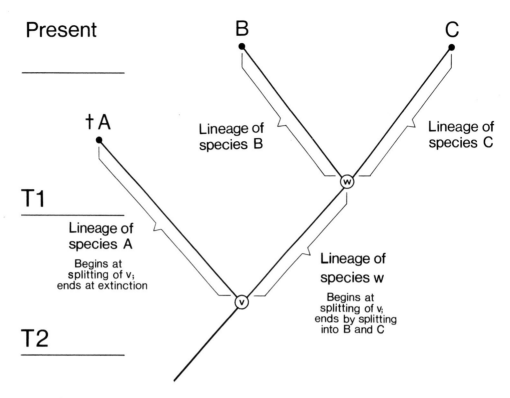

Fig. 3. Diagram of the phylogenetic kinship between the extinct species A and recent species B and C. Representation of the three possibilities of the existence of species in time. 1. Species A begins with the splitting of biopopulations of species v in the period T 2; it becomes extinct in T 1 without offspring. 2. Species w exists as a line of succeeding biopopulations between two processes of splitting (speciation). It begins, as species A does, with the splitting of biopopulations of the species v and ends with the dissolution of its last population into populations of the descendant species B and C. 3. B and C live as sister species in the present. Species v and w become stem species as a result of splitting, which the next diagram will depict in more detail. (Ax 1988)

gained rapid acceptance in the world of biology; I have also joined this school of thought (Ax 1984, 1987, 1988).

There are, however, a number of serious objections to the concept of "species as things", which are condensed in the alternative concept of "species as natural kinds" (BUNGE 1979; MAHNER 1993, 1994). A textbook of phylogenetic systematics must raise controversial ideas of a fundamental nature concerning the composition of life. However, we should, of course, always bear in mind that the daily work of applying the methodical procedure of phylogenetic systematics and in developing the system is not very concerned with whether species and descent communities are regarded as things or if they form natural kinds.

Objection to the "species as thing" concept is expressed in the following points:
– Binding relationships exist between the parts of a thing. This is not the case in allopatric populations of a species living in different places.
– The links between populations of a species in time do not go beyond the relations that actually exist between the individuals of a specific time.
– By combining certain parts, new attributes emerge in the resultant thing (material system) – the composition of an organism made up of different body parts, for example. This is not the case in the composition of species made up of individuals.

16

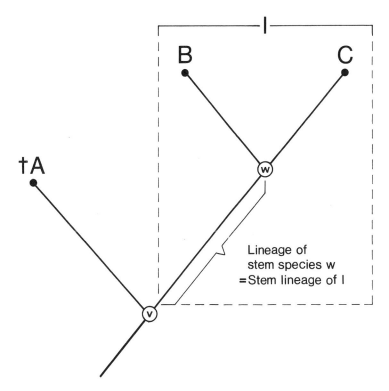

Fig. 4. The unity I is a descent community of minimal size. It is composed of the recent species B and C as well as the extinct species w; the latter becomes the stem species of B and C as a result of its dissolution into these sister species. The next-large descent community includes the species A, B and C as well as stem species, v, which is common to them alone. (Ax 1988)

– The parts of material systems can be things. This holds true for biological individuals as single organisms that live at the same time in a certain population. Material systems cannot, however, consist of things which no longer or do not yet exist (Mahner 1993). Let us take the biological individual Charles Darwin as an example. He was part of a certain population made up of humans living at the same time in the past century. He cannot, however, be a part of the present human population, cannot belong to a present material system. He is, nonetheless, a deceased member of the class *Homo sapiens*.

Does this mean that biological species are classes and, as such, merely arbitrary creations of our minds? The idea seems monstrous at first, but, as we will see, becomes less ludicrous on closer examination. In the context of class, philosophy makes use of the concept of natural kinds.

Natural kinds are not arbitrary collections of objects put together by man, but rather classes made up of real objects of Nature that share certain common attributes. These attributes are regularly connected in that they always appear correlated with each other. All members of a natural kind are nomologically equivalent. In the case of species, the individuals in space and time form the real objects.

The concept of the species as a natural kind is defined as follows (Bunge 1979, p.83; Mahner 1993, p.121):

A species is a biospecies if, and only if

1. it is a natural kind (rather than an arbitrary collection);

2. all of its members are organisms (present, past, or future);
3. it "descends" from some other natural kind (biotic or prebiotic).

I would like to emphasize two aspects of this definition.

– Species consist of individuals which are connected with each other through the succession of generations. The particular mode of reproduction occurring in the chain of generations must be ignored if all species are to be included in the definition. The modes of reproduction can thus include the common bisexual reproduction between individuals of a population with the reproductive isolation against individuals of other populations, unisexual, parthenogenetic reproduction as in the case of Bdelloidea (Rotatoria), or even partial asexual, vegetative propagation.
– Also, the definition leaves open the form of descent. Aside from the fact that the evolution to cellular organisms must be traced back to prebiotic material, the "origin of new species" is possible in different ways – through the splitting of biopopulations, but also through hybridization and polyploidy.

A word on the naming of species. The international rules of nomenclature prescribe a **binomial** for every species, comprising a genus name and an epitheticon. The four medusae from our example are thus called *Aurelia aurita* (common jellyfish), *Cyanea capillata* (red jelly) *Chrysaora hysoscella* (compass jellyfish) and *Rhizostoma octopus* (cauliflower jellyfish). Since the genus name is an obligatory component in the binomial, the assumption is often made that there are genera in Nature. This, however, is not the case. Categories such as genus, family, order, or phylum are labels of traditional classifications to rank supraspecific taxa; they have no place in the phylogenetic system (p. 20). For this reason, it is appropriate to speak of the first and second name instead of the genus and species name (Ax 1988).

Descent Community

The descent community now becomes relevant as a natural kind consisting of several species with certain, regularly correlated characteristics.

"A descent community consists of the individuals of species that have once-evolved, derived features which only they share. A descent community comprises at least one stem species and two descendant species." (Fig. 4).

In order to illustrate this point, I would like to refer again to the third level of our observations on jellyfish. Two features were detectable – the structure of the polyp with four septa, and the terminal budding of ephyra larvae, which only individuals of certain jellyfish species have in common. The theory of evolution explains these circumstances as follows. In the history of organisms there was once a jellyfish species in which the features mentioned were evolved for the first time as evolutionary novelties (autapomorphies). All currently existing jellyfish – species with four septa in the polyp and with ephyrs – originate from populations of this stem species.

In contrast to the profusion of nomological equivalences between populations of species, descent communities usually have only a few similarities which are manifested in all their members. Why is this so? According to our understanding of evolution, new descent communities cannot suddenly appear with an abundance of evolutionary novelties. They are far more likely to start with one apomorphy or very few new developments; and these novelties can, of course, change during the course of further evolution until they are no longer recognizable. This circumstance explains the difficulty in proving phylogenetic kinship, especially in geologically old unities.

On the other hand, descent communities can be delineated with impressive clarity, if recent members can be linked to a long stem lineage such as in the case of the Aves or the Mammalia (p. 43). Many evolutionary novelties can, for example, be detected in currently existing members of the descent community Mammalia – well-known features such as hair and mammary glands, anucleate erythrocytes, or the three auditory ossicles in the middle ear. These all arose in one stem lineage.

Descent communities are given **single names** to distinguish them from the binomials used to designate species. Just as the Mammalia discussed above are a descent community within the Vertebrata, our jellyfish form a descent community, Scyphozoa, within the Cnidaria.

Natural Kinds
as Taxa of the Phylogenetic System

Classification is the oldest term used to denote the results of man's attempts to establish a relationship between different objects on the basis of certain parameters. **Classification** also refers to the process that leads to this goal.

The objects of phylogenetic systematics are species and descent communities. The central aspect deals with the representation of the kinship relations that these have, as they originated in Nature. In interpreting species and descent communities as natural kinds, we are dealing with conceptional objects, and any attempt at portraying the phylogenetic order as it occurs in Nature is thus automatically a process of classification. This does not mean, however, that the resulting classification must represent a set of arbitrary choices. The phylogenetic system we are aiming at is a conceptional system that strictly follows the parameters set by nature. **Systematics** is, for this reason, the appropriate term for this discipline, and **systematization** an adequate circumscription of its activities (MAHNER and BUNGE, ms.).

The word **taxon** (pl. taxa) was introduced into the field of biology comparatively late (MEYER 1926; cf. MAYR 1982), but quickly became a key term. Today, every delimitable unit within any given classification is called a taxon. Phylogenetic systematics has strict conditions about the use of the term taxon, for which there are only two precisely defined meanings. We speak of the taxon species for the natural kind species, and of the supraspecific taxon for the natural kind descent community. When it comprises one stem species and all of its descendant species, the descent community becomes a monophylum within the system.

This reveals the path from Nature to the phylogenetic system. The system is the attempt to derive maximal natural equivalence classes from living Nature. The natural kinds are thus christened; they receive binomina or uninomina. We adopt them with their names into the system, in which they, as species taxa and supraspecific taxa, represent the basic elements.

Classes have definitions based on certain features. To the extent that species and descent communities are considered to be natural kinds and are therefore classes, the common features of their members do, indeed, form the basis for definitions.

It should again be emphasized that natural kinds are not arbitrary classes for which we can create definitions to suit our needs. It is Nature which determines the pattern of features of the natural kinds. Therefore, it appears suitable to me to talk not of a definition, but rather of a characterization by features of the natural kinds or of the corresponding taxa of the system.

Elimination of Categories

Categories were created by LINNÉ (1758), together with names for species and supraspecific unities. The four categories of his Systema Naturae – species, genus, order and class – have since expanded to about 20 commonly used terms in zoology.

It is important to distinguish clearly between taxa and categories to avoid the commonplace misuse that occurs even in modern textbooks. Taxa are unities of the system, categories are labels for taxa. Taxa are categorized in order to mark their rank in the hierarchy of classifications. It is precisely here, however, that categories are unusable in the phylogenetic system.

The **category species** has been prescribed by the international rules of nomenclature. The category species is an objective term insofar as it refers exclusively to the taxon species. The taxa of the system labeled with the category species do not, however, necessarily have one and the same rank. Only sister species occupy an identical hierarchical level in the phylogenetic system (p. 31). Let us take the example of the three famous Australian monotremes. The terrestrial species *Tachyglossus aculeatus* and *Zaglossus bruijni* are sister species; they are unified on the next higher hierarchical level in the supraspecific taxon Tachyglossidae (Fig. 7). Positioned opposite them on this level is the aquatic platypus *Ornithorhynchus anatinus*, the only recent species of Ornithorhynchidae.

The **categories for the supraspecific taxa** display correspondence only in the formal sense that agreement has been reached on the succession of categories from the lowest-ranking genus through family, order, and class (each with subcategories) to the high-ranking phylum.
However, (1) "nonarbitrary definitions for the supraspecific categories are not available" (MAYR 1969, p. 91); (2) there are no valid reasons whatsoever for assigning certain categories to certain taxa, for example like to the Insecta or the Arthropoda, the Mammalia, or the Vertebrata; and (3), for the supraspecific unities, too, there are always only two equally ranked taxa in one monophylum of the system as a result of dichotomous splitting (p. 26). Thus, only two subunities would be possible in the category "phylum" in the monophylum Metazoa; only two taxa could have the label "order" within the monophylum Insecta. The commonly found listings of 20-25 phyla of multicellular animals, of around 30 orders of placental mammals, or of comparable numbers of equally ranked insect unities, do not reflect the reality of living Nature.

The arbitrariness, and resulting impracticality, of the Linnaean categories for the characterization of the taxa in terms of rank cannot be better demonstrated than by examining a few selected textbooks.
GRUNER (1993) divides the animal "kingdom" into 25 phyla in KAESTNER'S TEXTBOOK OF SPECIAL ZOOLOGY (LEHRBUCH DER SPEZIELLEN ZOOLOGIE). The subkingdom "Protozoa" has one phylum termed Protozoa; apart from the highly questionable assignment of the same name to two differently ranking unities, "Protozoa" do not form a monophylum. In the subkingdom Metazoa the phyla 2–25 are sequentially numbered and all given the same rank.
In WURMBACH/SIEWING'S ZOOLOGY TEXTBOOK, VOLUME SYSTEMATICS (LEHRBUCH DER ZOOLOGIE. BAND SYSTEMATIK 1985), the subkingdom Protozoa is divided into 7 phyla, and the subkingdom Metazoa has only 5 phyla.
Can there be more fundamental differences? What is the student supposed to do with this divergent array of categories?
Other textbooks drop the use of categories partially, though in completely different directions. A

chapter ANIMAL PHYLA in WEHNER and GEHRING'S ZOOLOGY (ZOOLOGIE 1990), has 22 phyla. Besides the phyla – referred to as the highest categories – only genera and species – referred to as the lowest categories, are mentioned. In STORCH and WELSCH'S SYSTEMATIC ZOOLOGY (SYSTEMATISCHE ZOOLOGIE 1991), on the contrary classes and orders form the focal points of categorization. All taxa above the level of class are given letters or numbers; there are no phyla here at all. Simultaneously, however, in KÜKENTHAL'S MANUAL (KÜKENTHALS LEITFADEN 1993), STORCH and WELSCH divide the animals into 30 phyla, and what appears here as a phylum is presented there partially as a class.

A consistent phylogenetic system is possible only after all categories have been eliminated.

Establishing the Phylogenetic System

There is only one objective reference for mastering the diversity which exists in Nature – the kinship relations between species and descent communities which were created during phylogenesis through the splitting of biopopulations. They form the only Principium divisionis for an epistemologically flawless representation of the phylogenetic order in Nature. This also means that there can be only one system of organisms – the phylogenetic system.

How do we recognize phylogenetic relations, and how do we represent them adequately in the phylogenetic system?

Feature Patterns of Species
and
Ground Patterns of Descent Communities

Every species has a species-specific pattern of features. This pattern is the result of evolution; it is made up of early-developed, primitive features and later-evolved, derived features.

Every descent community has a certain ground pattern, and this, too, is a mosaic of original plesiomorphous features and derived apomorphous features. The ground pattern of every individual descent community corresponds to the pattern of features of the stem species which gave rise to the community by splitting.

The ground patterns of descent communities are of fundamental importance in any attempt at establishing the phylogenetic system. Since we do not know the stem species, the ground patterns of the communities must be deduced by comparing the feature patterns of their members (species), and be presented as hypotheses.

This brings us back to the feature patterns of species, which are the empirical sources we have at our disposal. By studying individuals from populations of recent and fossilized species, we are searching for evolutionary congruences between different species.

The Three Possibilities of Evolutionary Congruences

Four thousand species of mammals have, uniformly, a complicated sound-conducting apparatus with the three auditory ossicles malleus, incus, and stapes (Fig.9). The other terrestrial vertebrates have only one bone in the middle ear, which corresponds to the stapes of the mammals. The following hypothesis is used in the theory of evolution to explain these facts. In the phylogenesis of terrestrial vertebrates two new bones were integrated in the middle ear – the malleus and

22

the incus. At the turn of the Jurassic/Cretaceous period about 150 million years ago, a species existed in which a sound conducting apparatus with three auditory ossicles was realized as an evolutionary novelty. As a result of the splitting of this species, the new feature was passed down to the descendant species; the former species itself became the stem species of all currently living mammals, and the possession of three auditory ossicles became a derived feature in the ground pattern of the descent community, Mammalia. The common existence of three auditory ossicles and many other apomorphous features forces us to interpret 4000 species as members of one unity, Mammalia. As a result, we create one taxon for them in the phylogenetic system of the vertebrates.

So far, so good. The value of evolutionary congruences or similarities for the research of kinship and for establishing the phylogenetic system seems convincingly documented. That the issue is not quite so simple, however, is evident from the following example.

Some 1000 species of free-living Plathelminthes (flatworms) have an epidermis made up of ciliated cells. Their parasitic relatives, on the other hand – known as trematodes and tapeworms – are distinguished by a very different, non-ciliated skin. What was, then, more obvious than to use the evolutionary similarity "epidermis with cilia" and to unite the free-living species in one taxon, Turbellaria, which is still today the case in almost all classifications of flatworms. The presence of ciliation, however, does not help in determining phylogenetic kinship relations. Ciliated epidermal cells have existed since the Metazoa came into being; they were adapted into the feature pattern of the stem species of the Plathelminthes as an original feature and passed on to the free-living descendant species. In other words, they form a plesiomorphous feature in the ground pattern of the Plathelminthes. An evolutionary novelty, "ciliated epidermis", exists neither for the Plathelminthes as a whole nor for a part of the unity. The feature cannot be used to prove kinship, and a taxon "Turbellaria" cannot therefore exist in the phylogenetic system of the Plathelminthes.

Let us now move to the third possibility of evolutionary similarities, that of multiple independent evolution leading to a more or less similar appearance. The terrestrial Arachnida and Tracheata of the Arthropoda have similar tube-like excretory organs; they project out between mid-intestine and rectum into the body cavity. These organs are known as the Malpighian tubules in arachnids, myriapods, and insects. There is, nevertheless, a clear vote against their use in kinship research on this level of comparison. Based on the assessment of other features, the Arachnida are grouped together with the aquatic Xiphosura in the descent community Chelicerata, whereas the Tracheata are grouped with the aquatic Crustacea in the Mandibulata. Xiphosura and Crustacea have no Malpighian tubules. Established kinship hypotheses force the assumption that new, tube-like excretionary organs evolved twice, independently of each other, in the stem lineages of the Arachnida and the Tracheata, probably as a result of their separate transition onto land.

Phylogenetic systematics makes use of a clear terminology for the three types of evolutionary congruences (HENNIG 1950) – synapomorphy, symplesiomorphy, and convergence. I will explain the following definitions by means of a formal example in which A, B, and C (species or descent communities) form a unity which arose from the stem species v. Furthermore, we will place B and C with the stem species w as the closest possible phylogenetic relatives possible within this unity (Fig. 5).

Synapomorphy

"Congruence in a once evolved apomorphous feature between the closest possible phylogenetic relatives."

B and C both possess a feature which arose once in the lineage of the stem species, w – their common stem species. The evolutionary novelty is a unique, derived feature – an **autapomorphy** of a taxon comprising B + C.

Monotremata and Theria both possess a sound-conducting apparatus with three auditory ossicles in the middle ear, which evolved once in a stem lineage common only to them. This construction is a uniquely derived feature, an **autapomorphy** of the descent community Mammalia (Monotremata + Theria).

Symplesiomorphy

"Congruence in a plesiomorphous feature"

A and B have a feature in common which the stem species v of the descent community A + B + C already possessed.
Within the community, the feature corresponds in A and B only, because it transformed into a new state in the lineage of C.

Free-living Plathelminthes have ciliated epidermal cells which the stem species of all flatworms already possessed. Within the community Plathelminthes, epidermal ciliation corresponds in the free-living species only because a new, cilia-free skin evolved in the stem lineage of the parasites.

Convergence

"Congruence in a derived apomorphous feature with repeated, independent evolution."

A and C have a feature in common which appeared neither in the stem species v nor in the stem species w. The correspondence is the result of independent evolution of the feature with comparable manifestations in the lineages of A and C.

Arachnida and Tracheata have similar excretory organs in the form of Malpighian tubules which existed neither in the stem species of the Chelicerata nor in the stem species of the Mandibulata. The congruence is the result of the independent evolution of similar, tube-like excretory organs in the stem lineages of the Arachnida and Tracheata.

Sister Species – Sister Group
Dichotomy

Phylogenetic systematics continuously and consistently seek the closest possible phylogenetic relation between species and descent communities. This will be explained in detail with the help of examples (p. 31). Since this term has already been used in the definition of synapomorphy, we will now define its meaning precisely.

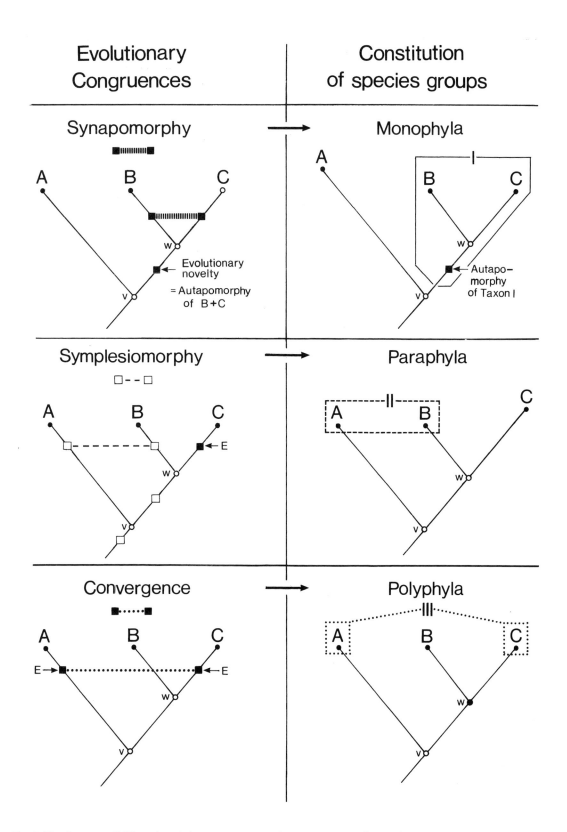

Fig. 5. The three possibilities of evolutionary congruence between two species or descent communities and the corresponding possibilities for establishing supraspecific taxa. As a prerequisite, we postulate an identical relationship between the unities A, B and C in all diagrams. E = evolutionary novelty. (After Ax 1988)

"The closest possible phylogenetic relatives are two species or descent communities with a stem species common to them alone. They are in the first degree of phylogenetic relationship with each other."

There are easily memorizable terms for this kinship relation. If two species are the closest possible relatives, then they are sister species. If two descent communities are in the first degree of phylogenetic relationship, they are called sister groups. It goes without saying that a single species can also be the sister species of a descent community.

In the search for sister species and sister groups, we presume the mode of dichotomous splitting of stem species into descendant species. This is a realistic, heuristic principle which is legitimized by the following facts and considerations:
– For a wide variety of different groups of organisms, it is often possible to base proof of sister species or sister groups on synapomorphous similarities common only to them. It can be argued, therefore, that dichotomous splitting represents a widespread mode of origin of species.
– The alternative of simultaneous multiple speciation of stem species into several descendant species is by far the more complicated process. The burden of proof lies with those who postulate such developments. If, however, a stem species did once actually fragment simultaneously into three, four, or more daughters, then phylogenetic systematics must treat this species group as a unity that methodically, cannot be further broken down (Ax 1984).

Lineage – Stem Lineage

Here I would like to explain the use of the terms lineage and stem lineage.
By the lineage of a species I mean the sum of its biopopulations which succeed each other in time. Evolutionary novelties can arise in the lineages of successive populations – as transformations of already present features or through the evolution of completely new features.
If a species becomes the stem species of a descent community through splitting, its lineage becomes the stem lineage of the community.

The Formation of Species Groups

The three forms of evolutionary congruence allow three possible species groups in classifications, i.e., three possibilities for the establishment of supraspecific taxa by man. We call them monophyla, paraphyla, and polyphyla (Fig. 5).

Monophyla

We create monophyletic taxa, monophyla for short, based on synapomorphous congruences. B and C are united to form monophylum I, and this corresponds to the origination from the stem species, w, common only to them. We adopt the sister species or sister groups B and C of Nature into the phylogenetic system as **adelphotaxa**.
We have termed evolutionary novelties from the lineage of the stem species w autapomorphies of the descent community B + C. We can carry this term on into the system and speak of the **aut-**

apomorphies of a monophyletic taxon, e.g., the middle ear with three auditory ossicles, as being an autapomorphy of the monophylum Mammalia.

Paraphyla

Species grouped together on the basis of plesiomorphous similarities do not have a stem species common only to them. In our example, A and B are united to form a paraphyletic taxon II. This occurs contrary to the phylogenetic relationships, according to which B and C are the closest relatives as adelphotaxa.

Traditional classifications feature numerous paraphyletic taxa, such as the "Turbellaria" mentioned above, the "Apterygota" as a taxon for the primarily wingless insects, the animals without a spinal column as "Invertebrata", the "Anamnia" as vertebrates without embryonic membranes, or the "Reptilia" within the Amniota.

The phylogenetic system does not have a place for the paraphyla; they must be phased out as quickly as possible. Until they can be entirely eliminated, their names should be written in quotation marks.

Polyphyla

The last possibility in our example is that A and C are united on the basis of independently developed evolutionary novelties to form taxon III. Polyphyletic taxa based on convergent agreements are generally rejected, and in my opinion there are no polyphyla today among the high-ranking taxa of the Metazoa. The "Protozoa" are, however, a mixture of eukaryotic organisms with primary unicellularity (plesiomorphy) and possibly multiple independently evolved heterotrophy (convergence; p. 47, 49).

The Change of Meaning
Synapomorphy – Symplesiomorphy

Evolutionary novelties (autapomorphies) can be synapomorphies only on one single level of the hierarchy, and that is in the comparison of the closest possible phylogenetic relatives. This is the level of sister species and sister groups in Nature, which have their equivalents in the phylogenetic system in the adelphotaxa.

However, as soon as we move down to subordinate unities within a monophyletic taxon, the meaningful synapomorphy turns into a worthless symplesiomorphy (Fig. 6).

Our example of the three auditory ossicles in the middle ear of mammals illustrates this point. We have postulated the evolution of this feature in the stem lineage of mammals and have interpreted its existence amongst all recent mammal species to be an autapomorphy of the monophylum Mammalia. On the level of their adelphotaxa Monotremata and Theria, the closest possible phylogenetic relatives of the Mammalia, the similarity in the construction of the middle ear hence forms a synapomorphy. However, if we compare single species of Monotremata or the Marsupialia and Placentalia as subordinate taxa of Theria with one another, the middle ear with three auditory ossicles proves to be a plesiomorphous similarity – a symplesiomorphy. This is because the feature did not evolve in the stem lineage of the Monotremata or the stem lineage of the Theria, but in the stem lineage of all recent Mammalia.

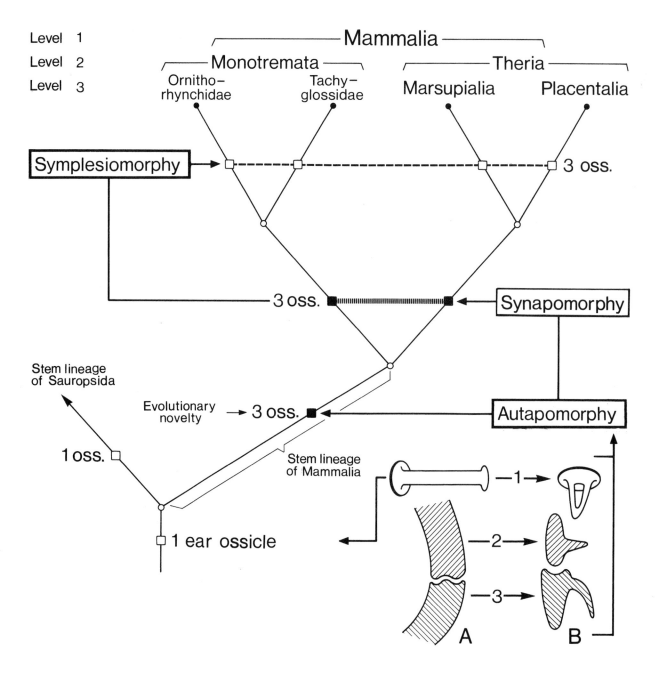

Fig. 6. Assessment of the feature "three auditory ossicles in the middle ear" on the upper three hierarchical levels of the phylogenetic system of Mammalia. The autapomorphy of level 1 (monophylum Mammalia) is a synapomorphy on level 2 (adelphotaxa Monotremata – Theria) and becomes a symplesiomorphy on level 3 (adelphotaxa Ornithorhynchidae – Tachyglossidae; adelphotaxa Marsupialia – Placentalia). The alternative "one to three auditory ossicles" is represented in the lower diagram on the right.

A. Plesiomorphous state in the Tetrapoda with one auditory ossicle (1), the primary jaw articulation composed of quadratum (2) and articulare (3).

B. Apomorphous state of the Mammalia with the auditory ossicles stapes (1), incus (2) and malleus (3). (Ax 1984)

Forming Hypotheses About Synapomorphy

Synapomorphy on the level of the closest possible phylogenetic relatives **is the decisive evolutionary congruence** for the study of kinship relations and for the establishment of the phylogenetic system. It is therefore our task to trace an epistemologically flawless path to propose hypotheses about synapomorphous congruences that are intersubjectively testable at any time. In my estimation, we must come to two decisions in two logically connected steps. First, it must be determined whether an empirically proven similarity is a plesiomorphy or an apomorphy. If we consider the latter to be probable we must then decide between a unique evolution or multiple independent evolutions of the similarity – the decision between synapomorphy and convergence.

Plesiomorphy – Apomorphy

The elementary and almost universally applicable method in this context is the **out-group comparison**. After having detected an evolutionary similarity between certain species, I first look for an alternative. This can be a different manifestation of the feature, but also its absence. What is original here, and what is derived? A decision based on probability can be made by a comparison with the more distant "relatives", whereby the out-group does not need to be more closely circumscribed. The following applies for the conditions just outlined:

"If different states of a feature appear in a species group, it is highly probable that the manifestation which is realized in the more distant relatives – in the out-group – is the plesiomorphy."

As opposed to the three auditory ossicles in the 4000 species of mammals, we find only one sound-conducting bone in the middle ear of lizards and snakes, crocodiles and birds. This has already been mentioned. Within the unity Amniota – all vertebrates with the embryonic membranes amnion and serosa – there is, in other words, the alternative of one to three auditory ossicles. Which is the plesiomorphous state, which apomorphy? As the more distant four-footed terrestrial vertebrates, the Amphibia can serve as the out-group. They have one auditory ossicle, and so the decision is made on the basis of the following hypothesis. The existence of one auditory ossicle (stapes) is the plesiomorphy; the manifestation of three auditory ossicles (stapes, malleus, and incus) is the apomorphy of the alternative.

The Aphaniptera (fleas) are wingless ectoparasites of birds and mammals. They are grouped with the winged beetles, butterflies, hymenopterans, and many other insect unities in the huge descent community of the Holometabola, characterized by a pupal stage during which there is no food intake. The assessment of the alternative "presence or absence of wings" can again be concluded by means of the out-group comparison. Two pairs of wings on the thorax, for example amongst the libellans, grasshoppers, or bugs, are common even outside the group of holometabolic insects. We thus conclude that within the Holometabola, the fleas' absence of wings is the apomorphy of the alternative.

Plausible explanations of the constructive-functional change of features and estimates of their adaptive state supplement the out-group comparison (Ax 1988).

Synapomorphy – Convergence

Did the evolutionary change of the sound-conducting apparatus from one to three auditory ossicles come about once within the stem lineage of the Mammalia, or perhaps several times convergently in different subgroups of the mammals? Was there once a wingless stem species of all present-day fleas, or were the wings reduced within different unities of the Aphaniptera independently of one another?

We have no definite answers to these questions, as there is no empirical measure for the decision between synapomorphy and convergence.

We do not know of any criterion, nor do we have any tests on the basis of which we could decisively tell if a certain evolutionary similarity between two species (groups) developed once or several times. It is worth mentioning here that it is in principle irrelevant to our discussion whether the congruence is absolutely identical or if there are differences in detail. Nature solved the problem long before our time for each individual case. The human cognitive apparatus is, at best, capable of offering rational answers. The **Principle of Parsimony** (Ockham's Razor) – the principle of the simplest or most parsimonious explanation – suggested by the philosopher William of OCKHAM in the 14th century is relevant here. Our propositions will have a scientific character only if they offer the simplest and most parsimonious explanations for our empirical findings.

The hypothesis of a unique evolution of features which appear as congruences between different species or descent communities is decidedly more parsimonious than the hypothesis of multiple independent evolutions leading to the same result.

The principle of parsimony dictates the following answers to the questions posed above. The integration of two new bones in the sound-conducting apparatus of the middle ear occurred once in the stem lineage of the Mammalia. The reduction of wings in the Aphaniptera occurred once on the transition to ectoparasitism.

These are and remain, of course, hypotheses. They are reinforced or confirmed when they match the assessment of other features of the unities in question – when, for example, the interpretation of the auditory ossicles does not conflict with the assessment of numerous other features as synapomorphies of the Monotremata and Theria.

On the other hand, we do not have a principle, and there is no legitimation for postulating a multiple, independent evolution of similar features a priori. Conflicting "evidence" may, however, very well require a posteriori the assumption of convergence for individual similarities. We have already been able to show this in the evolution of the Malpighian tubules in the Arachnida and Tracheata because of their integration in different descent communities.

The following objection to this argumentation is often made. We do not know anything about the possible complexity of evolutionary processes. Why should Nature not have brought forth a certain result repeatedly and independently? Nature is evidently in a position to do this, as the example of the Malpighian tubules shows, but this is not the focus of our attention in our demand for the most parsimonious interpretation possible. Using the principle of parsimony as a methodological instrument does not presume that Nature always took the evolutionary path that appears the most simple to our way of thinking. Nature may well have shaped her processes as extravagantly and complicatedly as she was "willed" to do. We, on the other hand, have no other choice than to strive for the most parsimonious explanation if we want to interpret the results of evolution in a rational way and represent them in justified kinship hypotheses. Ignoring this requirement would mean losing all claim to the scientific character of our endeavors.

Homology

"Homology of a feature in different species or descent communities depends on the adoption of the feature from a common stem species" (Ax 1988, 1989a).

This formulation defines the meaning of the word homology which was the main term in the study of phylogenetic relationships (Remane 1952) until the terminology of phylogenetic systematics was developed. Within the framework of this terminology, two aspects of the term homology should be emphasized:

– Homology covers both synapomorphy and symplesiomorphy. It does not differentiate between these two fundamental forms of evolutionary congruence.
A homologization of the three auditory ossicles, malleus, incus, and stapes in the middle ear, and of the humerus in the upper arm of the chimpanzee and of man, will hardly meet with much opposition these days, except perhaps from advocates of a theory of Creation; but on this low-ranking level the homologies are meaningless symplesiomorphies. It is only when we reach the relevant level in the hierarchy of the adelphotaxa that homology becomes the decisive synapomorphy for the study of kinship relations – in the case of the auditory ossicles on the level of the mammalian sister groups Monotremata and Theria, and in the case of the humerus on the level of the tetrapod sister groups Amphibia and Amniota (p. 36, 37).
– Homology cannot terminologically cover the absence of features, and therefore cannot realize the value of this absence for the study of kinship relations. It is not possible to homologize anything that is not present. This applies, for example, to the primary absence of wings in the apterygote insect unities Diplura, Protura, or Collembola (symplesiomorphy) and their secondary absence in the sister groups of the Aphaniptera (synapomorphy).

In the above-mentioned assessment of evolutionary similarities, we started with a decision between plesiomorphy and apomorphy. This is nothing but a method that is recommended. It might be worthwhile in individual cases to first formulate a homology hypothesis for a conspicuous similarity between different species, and then to look for the relevant level on which it should be decided between synapomorphy and convergence.

Whichever the path chosen, we have to emphasize the point already mentioned about the probability of similarities as possible synapomorphies when applying the principle of parsimony. There are no methodical criteria with which homologies can be identified, recognized, or determined. Homologizations of features between different species or descent communities are and remain hypotheses on random levels of comparison.

The Search
for the Sister Species or Sister Group
and the Resulting System

The central focus in the study of kinship relations is the ongoing search for the closest possible relatives of species and descent communities. The following takes a look at the work involved in an examination of the terrestrial vertebrates, the unity Tetrapoda.
The resulting system brings sister species and sister groups onto one and the same level in

accordance with the simultaneous origin from the dichotomous splitting of their stem species (p. 26). This leads us to the two possibilities for an identical representation of kinship relations which we spoke about in the introduction:

– the hierarchical tabulation of the system with the indentation of adelphotaxa pairs corresponding to the order of their subordination.
– the relationship diagram with these pairs arranged in successive levels.

Hierarchical Tabulation

Tetrapoda
  Amphibia
  Amniota
    Sauropsida
    Mammalia
      Theria
      Monotremata
        Ornithorhynchidae
        Tachyglossidae
          Tachyglossus aculeatus
          Zaglossus bruijni

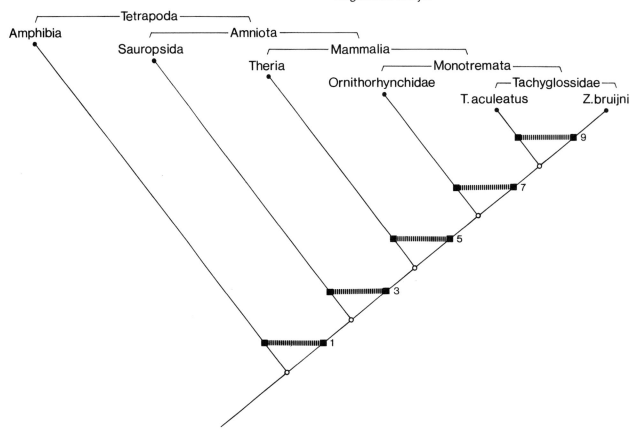

Fig. 7. Diagram of phylogenetic kinship relations within the Tetrapoda. The synapomorphies listed justify the five adelphotaxa relations Amphibia – Amniota (1), Sauroposida – Mammalia (3), Theria – Monotremata (5), Ornithorhynchidae – Tachyglossidae (7) and *Tachyglossus aculeatus* – *Zaglossus bruijni* (9)

32

Relationship Diagram

The relationship diagram is reproduced twice. The synapomorphies 1,3,5,7,9 of the five pairs of sister species/sister groups are listed on one side; these justify the five adelphotaxa relations which are postulated (Fig. 7). Their respective autapomorphies 1,3,5,7,9 are listed on the other side in the stem lineages of the descent communities which enclose them; they developed as evolutionary novelties in these stem lineages (Fig. 8). The autapomorphies 2,4,6, and 8 have been added to justify all supraspecific taxa named as monophyla.

The Methodical Path

We will now outline step by step the search for sister species/sister groups. An overview of the features mentioned is given subsequently.

Let us begin on the lowest hierarchical level with the sister species *Tachyglossus aculeatus* and *Zaglossus bruijni* from the Australian region (Fig. 9). The echidnas are original terrestrial animals (plesiomorphy) with long defensive spines (apomorphy) and claws for cleaning them (apomorphy). Applying the principle of parsimony, we assess the apomorphies to be products of a unique evolution. In this interpretation they become synapomorphies of the two terrestrial species and autapomorphies of the unity Tachyglossidae (9) in which they are included.

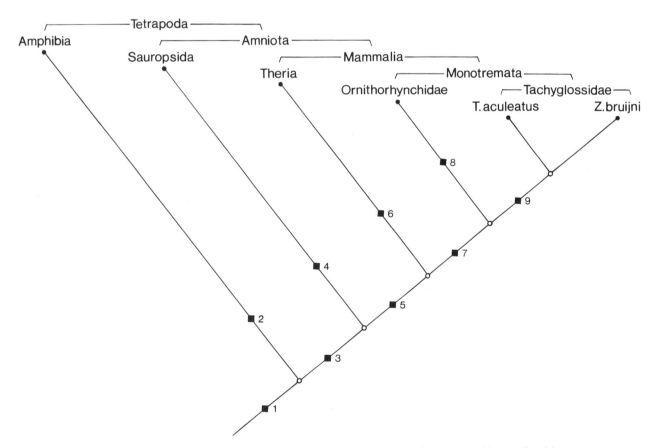

Fig. 8. Identical diagram of kinship relations within the Tetrapoda. Now nine groups of autapomorphies are listed here which originated as evolutionary novelties in the stem lineages of the Tetrapoda (1), the Amphibia (2), the Amniota (3), the Sauropsida (4), the Mammalia (5), the Theria (6), the Monotremata (7), the Ornithorhynchidae (8), and the Tachyglossidae (9)

The sister group of the Tachyglossidae are the Ornithorhynchidae with the recent platypus *Ornithorhynchus anatinus* and a fossil species from the Miocene epoch.

Why do we unite these two unities under the name Monotremata? The males of the three species have a unique organ in the posterior extremities which is not known in any other mammal. This organ is a large crural gland and a keratinous spur next to the five toes used to inject the glandular venom into other animals (Fig. 9). I will repeat the argumentation once again. The principle of parsimony dictates the assumption of a unique evolution of the crural gland and spur in a stem lineage common only to the Tachyglossidae and Ornithorhynchidae – and this assumption does not conflict with other facts. Therefore, the apomorphous feature becomes a synapomorphy of the Tachyglossidae and Ornithorhynchidae, and an autapomorphy of the unity Monotremata (7).

It is worth listing the special features of *Ornithorhynchus anatinus* (8). This species lives mainly in freshwater (apomorphy). A whole complex of conspicuous adaptations are connected with the change of biotope – a streamlined body, webbed feet, a beaver-like tail, a horny duck-like bill for dabbling (Fig. 9); but compared to the sister group Tachyglossidae, *Ornithorhynchus anatinus* with its soft coat of hair (as opposed to spines) has retained a plesiomorphy.

In the next step, we look for the sister group of the egg-laying monotremes, which possess a cloaca and have milk flowing diffusely out of mammary areas. I have just named three famous plesiomorphies, the alternatives to which are found in all other mammals. The separation of the anus and urogenital system, the birth of living offspring, and the termination of mammary ducts in teats can be interpreted, without any contrary indications, as autapomorphies of the Theria (6) i.e., of all mammals with the exception of the three species of monotremes.

What, however, are the reasons for unifying the Theria and the Monotremata into one monophylum, Mammalia? There are more than 60 features which are assessable as apomorphies compatible with each other, all of which must have developed in the stem lineage of the Mammalia. There are only a few unities of the phylogenetic system of the Metazoa that can be so clearly justified as Monophyla as the Mammalia. We have compiled some prominent synapomorphies of the Monotremata and Theria below (5), including, of course, the middle ear with three auditory ossicles, which has been quoted so often by now that this is the last time it will be mentioned.

Far more unsatisfactory is the search for the sister group of the mammals. The Sauropsida, which has often been considered to be a unity of "reptiles" and birds, remains a debatable candidate; special features in the circulatory system or in the development of a joint between the tarsi of the posterior extremity are but poor indications of possible autapomorphies (4). In recent literature, an adelphotaxa connection between the Aves and Mammalia has been vehemently suggested (GARDINER 1993).

Whatever the case may be, the entirety of "reptiles", birds and mammals can be unhesitatingly justified as a monophylum Amniota. Of the autapomorphies (3), I will draw attention to the laying of hard-shelled eggs on land – eggs with a microaquarium for the development of embryos through the formation of the embryonic membranes amnion and serosa.

In the final step, we search for the sister group of the Amniota among the "terrestrial" vertebrates. This are the Amphibia, which show distinctive plesiomorphous characteristics with their extensive bonds to the aquatic environment – egg-laying, aquatic larvae, etc.; but the stem species of the Amphibia also possessed special features (2). The pentadactyl extremities had barely evolved as the prominent autapomorphy of the Tetrapoda (1), when the fifth digit was reduced in the stem lineage of the Amphibia, as can be seen in all water newts and frogs.

The next step would be to search for the sister group of the Tetrapoda among the fish-shaped vertebrates; the Dipnoi (lung-fish) and *Latimeria chalumnae* (Actinistia) are hotly debated candidates.

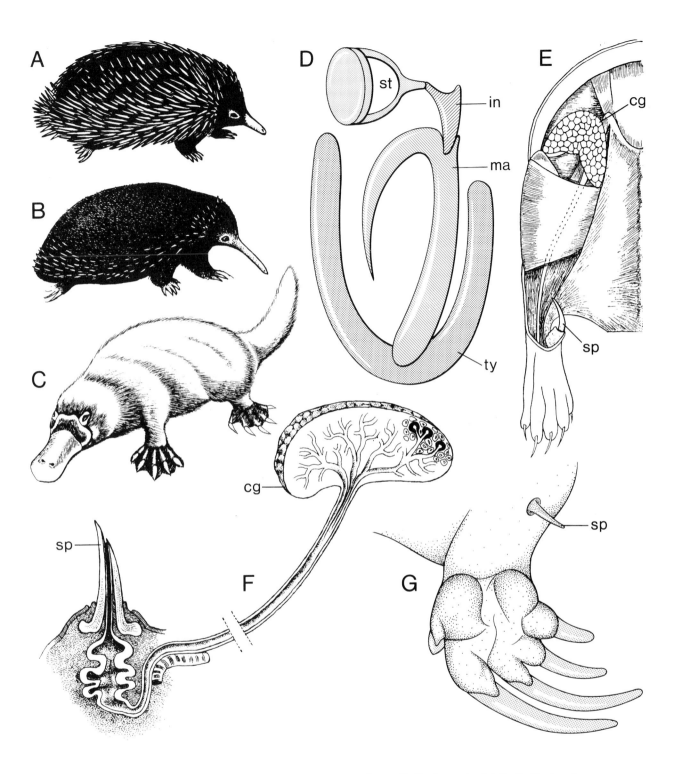

Fig. 9. A-C. Recent species of the Monotremata. A. *Tachyglossus aculeatus.* B. *Zaglossus bruijni.* C. *Ornithorhynchus anatinu*s. D. Original state of the three sound-conducting bones of the middle ear malleus, incus and stapes as well as the tympanicum in the Theria (Mammalia). E. Posterior extremity of a male from *Ornithorhynchus anatinus* with crural gland and spur. F. Dissection of the crural gland of *Ornithorhynchus anatinus.* longitudinal section of spur and secretory reservoir. G. Posterior extremity of *Tachyglossus aculeatus* with long claws on the second and third toe for cleaning. The spur is also visible. cg = crural gland. in = incus. ma = malleus. sp = spur. st = stapes. ty = tympanicum. (Various authors, in Ax 1988)

We will stop here, however, and end with an overview of conspicuous autapomorphies and synapomorphies of the taxa we have been discussing. Before doing so, I would like to mention one more important fact which played a role in the preceding argumentation without being specifically mentioned. If we want to justify two species groups as adelphotaxa, then alternatingly for both unities, features must be hypothesized as autapomorphies, each of which has respective plesiomorphies in the partner.

Justification of the discussed unities as monophyla (numbers in brackets refer to numbers in Fig. 8).

■ **Autapomorphies (9) of the Tachyglossidae**
= Synapomorphies (9) of *Tachyglossus aculeatus* and *Zaglossus bruijni*.
− Spiny coat.
− Special claws on the hindlimbs for cleaning the body covering.

■ **Autapomorphies (8) of the Ornithorhynchidae**
− Semiaquatic life-style.
− Streamlined body.
− Duck-like bill.
− Horizontal, flattened tail for steering.
− Webs on the forelimbs and hindlimbs.

■ **Autapomorphy (7) of the Monotremata**
= Synapomorphy (7) of the Tachyglossidae and Ornithorhynchidae.
− Crural gland and spur on the hindlimbs of the male.

■ **Autapomorphies (6) of the Theria**
− Separation of anus and urogenital system.
− Vivipary.
− Teats.

■ **Autapomorphies (5) of the Mammalia**
= Synapomorphies (5) of the Monotremata and Theria.
− Synapsid skull.
− Secondary jaw articulation between squamosum and dentale.
− Middle ear with three auditory ossicles and tympanicum.
− Heterodont dentition.
− Diphyodont tooth replacement.
− Homoiothermy.
− Hair.
− Anucleate erythrocytes.
− Sweat glands, sebaceous glands, mammary glands.
− Secondary palate.
− Number of phalangal joints 23333.
− Rotation of the extremities under the body.

■ **Autapomorphies (4) of the Sauropsida**
– Mesotarsal articulation.
– Hook-shaped metatarsal bone 5.

■ **Autapomorphies (3) of the Amniota**
= Synapomorphies (3) of the Sauropsida and Mammalia.
– Thick, horny skin. Formation of horny scales and claws.
– Embryonic membranes amnion and serosa. Allantois.
– Direct development. No aquatic larvae.
– Atlas and epistropheus (axis) as differentiations of the first two cervical vertebrae.

■ **Autapomorphies (2) of the Amphibia**
– Anterior extremities have only the digits 1-4.
– Occipital bone with double condyle.
– Teeth divided into crown and pedicel.
– Papilla amphibiorum (sensory termination in the inner ear).

■ **Autapomorphies (1) of the Tetrapoda**
= Synapomorphies of the Amphibia and Amniota.
– Change of biotope water/land.
– Five-toed forelimbs and hindlimbs. (Extremities originally spread from the body).
– Primary number of phalangal joints 23454.
– Lung as the only respiratory organ of adult.
– Choana and nasolacrimal duct.
– Middle ear with eardrum and one auditory ossicle (stapes).
– Uniform vertebral bodies with interconnections (zygapophyses).
– Unpaired occipital condylus on the skull ventral to the foramen magnum.

In the example of the Tetrapoda, we have written the system in accordance with the continuing search for the sister group, starting from the lowest-ranking level of two species and moving up to increasingly higher hierarchical levels.

When representing the system of the Metazoa, we are obviously following the reverse of the order in this book. We are taking the results of the search for sister groups as a prerequisite, and are developing a representation from them which proceeds step by step from the high-ranking to the low-ranking hierarchical levels.

Direct Access to the Autapomorphy

As we have shown, synapomorphies between sister species and sister groups automatically form autapomorphies in the ground pattern of the descent community which they comprise. **Autapomorphies are the decisive features with which we justify hypotheses about the existence of descent communities in Nature.**

Deliberations on the simplest possible way of finding autapomorphies should be of particular interest in this context. In order to identify a certain feature as an autapomorphy of a species group, it is not absolutely necessary to know the sisters (adelphotaxa) in which the similarity appears as a synapomorphy. This is always the case when a clearly apomorphous feature is realized in all species of the species group in question. The following example is representative of an abundance of similar cases in which the argumentation is clear.

Mosquitoes and flies have only one pair of wings, in contrast to all other winged insects. In the impressive number of 85000 species, there are two tiny knobs in place of the second pair of wings. They are called halteres. The principle of parsimony allows us to formulate the following hypothesis: in the phylogenesis of the Insecta, the rear pair of wings was once transformed into minute halteres; an insect species existed about 250 million years ago in which halteres were realized as an evolutionary novelty. Upon splitting, it became the stem species of all mosquitoes and flies – and the halteres became a prominent autapomorphy of a huge descent community Diptera. In order to justify the Diptera as a monophyletic species group, we do not need to show proof of the fossil sister species which developed from the splitting of the stem species, nor do we need to know the two highest ranking sister groups of the recent Diptera, regardless of whether they are mosquitoes and flies or not.

Molecules and Computers

After the electron microscope revealed an immense new field of ultrastructural features, two disciplines emerged as the latest players in the area of phylogenetic research – molecular biology and computer analysis.

"Molecular Systematics"

HILLIS and MORITZ (1990)

"Because the organisms under study have a single history, systematic studies of any set of genetically determined characters should be congruent with other such studies based on different sets of characters in the same organism" (HILLIS 1987, p.23).
"Each organism's phenotype and its underlying genotype have experienced the same evolutionary history, hence, both general types of data sets should provide the same estimates of phylogenetic relationships among species." (MEYER 1993, p.131).

The appearance of the individual organism is a result of the expression of its genetic information. Consequently, phenotypical similarities in different individuals should be the expression of identical genes, and nonmodificatory differences between them should trace back to mutations in the nucleotide sequences of DNA. There are often large discrepancies, however, which contradict the assumption of congruence between morphology and molecules (PATTERSON et al. 1993; RAFF et al. 1994; WÄGELE 1994a; WÄGELE and WETZEL 1994). These discrepancies apply especially to the assessment of kinship relations of geologically old, high-ranking unities, such as are the subject of this book. Contradictions arise not only due to the contrasts between structural and molecular features; there are also contradictions within molecular biology in the assessment of molecular feature patterns.

In a highly regarded study MOLECULAR PHYLOGENY OF THE ANIMAL KINGDOM, FIELD et al.(1988) arrive at the hypothesis of a "polyphyletic origin of Metazoa", based on sequencing the 18S ribosomal RNA of individual species. In their opinion, Cnidaria and Bilateria evolved independently of each other from different protists. The logical requirement of phylogenetic systematics for designation of their respective sister groups in the unicellular organisms is not considered. The hypothesis of an independent development of Cnidaria and Bilateria contradicts a number of common, derived features in their morphology and development. These include the congruence of ectoderm and endoderm, and the presence of sensory cells, nerve cells and muscle cells, all of which they have in common. More than a dozen autapomorphies indicate clearly that the Metazoa, including the Porifera and Placozoa, which were not mentioned by FIELD et al., form a monophylum (p. 54).

However, the hypothesis of a "double origin of the Metazoa" also finds "molecular support" (CHRISTEN et al. 1991). On the basis of sequence analyses of the 28S rRNA, the authors presume an independent aggregation of unicellular organisms in two different lineages, which supposedly lead to the "Diploblastica" (Porifera, Cnidaria, Ctenophora) and to the Triploblastica (Bilateria). "If such an early separation between diploblasts and triploblasts is confirmed by later analyses, the monophyletic origin of metazoa will be extremely difficult to demonstrate beyond doubt" (CHRISTEN 1994, p.471). This claim suggests that "molecular phylogenies" are the infallible answer to conflicting phylogenetic kinship hypotheses.

This claim is, however, relativized by molecular biology itself. LAKE (1990) analyzed anew the sequence data of the 18S rRNA of FIELD et al. (1988), and categorically stated "the Metazoa is monophyletic". This hypothesis receives "molecular support" with the help of sequence analyses of the srRNA (VAN DE PEER et al. 1993) and 16-like rRNA (WAINRIGHT et al. 1993), for example, or in similarities between the amino acid sequences of extracellular adhesion molecules, of adhesion receptors and nuclear receptors of different metazoans (MÜLLER et al. 1995).

Yet another variant may shed some light on the multifaceted nature of "molecular phylogenies" when dealing with the basal branching of the Metazoa. LAFAY et al. (1992) proclaim a large genetic distance and a correspondingly deep cleavage between three groups of sponge species which belong to the Calcarea and Demospongea in the traditional classification of the Porifera. On the basis of sequences of partial 28S rRNA, they claim that certain sponge species groups are more closely related to the Placozoa, Cnidaria and Ctenophora than to one another. I have deduced the following, partially contradictory, hypotheses from three diagrams of the work in question:

- The Demospongea species *Petrosia ficiformis* and *Reniera mucosa* form the sister group of the Placozoa + Cnidaria + Ctenophora and of all other Demospongea and Calcarea. Hexactinellida were not studied.
- The Demospongea species *Crambe crambe*, *Dictyonella incisa*, *Petrosia ficiformis*, and *Reniera mucosa* together are the adelphotaxon of *Trichoplax adhaerens* (Placozoa).
- The Calcarea species *Clathrina cerebrum* and *Petrobiona massiliana* are in one diagram closely related to the Ctenophora, and in another they are the sister group of the Cnidaria.

What conclusions can we draw from these ideas?

- The Porifera are not a monophylum, despite a number of apomorphous congruences in morphology common only to them (p. 68).
- The Porifera are a paraphylum. The Placozoa, Cnidaria, Ctenophora, and also the Bilateria have developed separately from each other from different stem lineages of sponges.
- The Porifera may even represent a polyphylum. Heterotrophic multicellular organisms which are organized as sponges would then have developed several times independently as the respective sister groups of the Placozoa, Cnidaria, and Ctenophora.

These are all possible ways of thinking while sitting at one's desk. They are all incompatible, however, with the phylogenetic systematization of the Metazoa as developed in this book. The Porifera can be justified as a monophylum without any qualifications; they form the sister group of all other metazoans unified under the name Epitheliozoa (p. 67).

Let us leave the "...confusion surrounding molecular systematic studies of animal origins..." (SOGIN 1994, p.184) and consider some ideas of a general nature. I have grouped them into four main points.

1. The comparison of molecular sequences has the unquestionable advantage of being quantifiable. As for the rest, the identification of congruences in the nucleotide sequences of DNA and RNA, in the sequence of amino acids in proteins, or in the molecular pattern of different organisms in general, is a priori nothing more than a demonstration of similarities. Hence, they per se have no relevance in the justification of phylogenetic kinship (WÄGELE and WETZEL 1994), just as is the case with unassessed similarities in structural features. Before molecular similarities can be used in the study of kinship relations, one must differentiate between plesiomorphous and apomorphous congruences. Furthermore, a decision as to whether apomorphies are synapomorphies or convergences must have been taken based on the principle of parsimony.

An example of a corresponding methodical procedure on the level of molecules is the assessment of two structures of the mitochondrial DNA in the monophylum, Cnidaria. According to a representative analysis of 48 species, the Anthozoa have, without exception, a ring-shaped mtDNA molecule. The Hydrozoa, Cubozoa, and Scyphozoa, on the other hand, are distinguished by a linear structure of the mtDNA (BRIDGE et al. 1992). By means of the out-group comparison, the common, ring-shaped molecule found in the Eucaryota is easily hypothesized as a plesiomorphy, which means that the linear formation in the Hydrozoa, Cubozoa, and Scyphozoa can be interpreted as the apomorphy of the alternative.

In the case of differences in the ultrastructure of the mechanoreceptor of nematocyst cells, the findings match this assessment. In the Anthozoa, the cnidocyte bears a "normal", motile cilium with a striated rootlet and an accessory centriole. This is undoubtedly an original state. The apomorphous alternative is found in the Hydrozoa, Cubozoa, and Scyphozoa; their nematocyst cells have a complex, stiff cnidocil without a ciliary rootlet and without an additional centriole (Fig. 27). According to the principle of parsimony, the apomorphous manifestations of a molecular pattern and of a structural feature can in the next step be hypothesized as synapomorphies of the sister groups Hydrozoa and Rhopaliophora (Cubozoa + Scyphozoa) and there is no conflict here. The linear conformation of the mtDNA and the cnidocil of the nematocyst cells likewise become autapomorphies in the ground pattern of the monophylum Tesserazoa (Hydrozoa + Rhopaliophora) within the Cnidaria (p. 87, 92).

2. Mention of the ground pattern brings me to the next point. "Molecular phylogenies" are, as a rule, based on the analysis of molecular patterns of single species of high-ranking taxa. The species are understandably chosen due to their easy availability, which does not, however, mean that they are primarily distinguished by original traits. The opposite is the case in the highly popular *Hydra* species of the Cnidaria, in the limnetic planarian *Dugesia tigrina* among the Plathelminthes (p. 141), and in the earthworms as representatives of the Annelida. We must, however, work our way to the ground pattern of monophyletic taxa, because only the comparison of ground pattern features can lead to satisfactory hypotheses in the study of kinship relations. In other words, a significant number of species of the diverse subtaxa should be analyzed for each monophylum in order to be able to establish the sequences of the 18S rRNA or 28S rRNA in the ground pattern of the Cnidaria, Plathelminthes, Annelida, etc. Precisely this occurs while working out the "ancestral sequence for the latest common ancestor of the living echinoid taxa" (SMITH 1992, p.228). "A robust molecular answer" to disputed questions about kinship relations within the Echinodermata is expected from the comparison with corresponding ground pattern sequences of the 28S rRNA of other high-ranking taxa.

3. Even a robust molecular answer, however, does not provide us with a final decision, and it cannot serve as the yardstick in the assessment of kinship hypotheses which phylogenetic systematics has established on the basis of other features. The commonly held high opinion of the value of molecular data as an "independent test" or standard per se is expressed in the following phrases: "Sequence data obtained from ribosomal RNAs offer an important new source of informative characters for inferring high-level phylogenetic relationships for many taxa and provide an independent test of hypotheses based on morphological characters, especially for metazoans" (TURBEVILLE et al. 1992, p.236). "Molecular characters have the potential to test monophyly and the internal phylogeny independently of morphological, developmental and ultrastructural characters" (RIUTORT et al. 1993, p.71).

This position cannot, I feel, go without question. All available features are equal in the analysis of kinship relations; they have no "hierarchy". No feature decides a priori the meaning of other features. Every manifestation of any given feature is significant if there is an undisputed assessment as autapomorphy together with corresponding interpretations of other features.

4. Extensive molecular biological research begins with single species of high-ranking taxa. It is also understandable that only a few taxa can, at first, be chosen from the entire spectrum of unities such as Metazoa, Bilateria, Spiralia, Deuterostomia, Vertebrata and others. The method becomes unacceptable, however, when the selected taxa are brought together in a relationship diagram. MÜLLER et al. (1995) suggest that with the Porifera (*Geodia cydonium*) and Nematoda (*Caenorhabditis sp.*) two "invertebrates" and four vertebrates arise from one root. Consistent interpretations of the branching lead to the conclusion that the Nematoda are the sister group of the Vertebrata.

As optimists, we should share the expectation "that when considering the logic of phylogenetic systematics and when carefully employing very long sequences ... a greater correlation will be obtained between the results of the sequence analyses and the studies of informative morphological features" (WÄGELE 1994a, p.231).

"Computer Cladistics"

Today, apart from the manual analysis and evaluation of features, it is possible to process extensive data using computers. Computer-supported data analyses do not, however, introduce fundamentally new elements into the methodology of phylogenetic systematics. Most recent surveys of the use of computers in phylogenetic analyses (MEIER 1992) and of methodical problems of "computer cladistics" (WÄGELE 1994b) present correspondingly differentiated assessments of their importance.

MEIER regards the computer as an indispensable tool for clarifying kinship relations on a low hierarchical level. On the other hand, when working with feature alternatives of an unproblematic "polarity" (plesiomorphy – apomorphy) and with few conflicting features, manual analysis is easily possible (MEIER 1992, p.107). In my opinion, this is true especially with respect to the search for sister group relations between high-ranking, geologically old taxa, for which there are often only few similarities available that can be interpreted as synapomorphies (BRUSCA and BRUSCA 1990, p.881; cf. p. 18).

WÄGELE (1994b) specifies the requirements of phylogenetic systematics which go to form the prerequisites for the creation and evaluation of a data matrix. The monophyly must be justified by autapomorphies and the ground pattern deduced for all taxa before they are incorporated into the computer analysis. Only features of the ground pattern are suitable for the data matrix. It must also be clear which of the congruences in the ground patterns of different monophyletic taxa can be hypothesized to be synapomorphies. When considering these requirements, it should be possible to resolve the fundamental differences between the manually generated kinship analyses conducted in this book and the computer analyses done, for example, by MEGLITSCH et al. (1991) or by EERNISSE et al. (1992), which are often widely contradictory themselves. "The belief of many users of computer programs, that they can objectively find sister group relationships and monophyletic groups without a priori assumptions, is a dream" (WÄGELE 1994b, p.104).

Terminology

An essential basis of understanding in every science is the use of a precise terminology. The following observations are meant to clarify and should not be interpreted as carping.

In the section on molecules, the alternative "polyphyletic origin – monophyletic origin of Metazoa" was discussed. The terms monophylum and polyphylum refer, however, to the composition of taxa (p. 26) and not to their origin. Furthermore, polyphyla are artificial products of man – they could not have originated in Nature at all. In other words, no valid unity of the phylogenetic system can have a polyphyletic origin. If the heterotrophic multicellular organisms were actually a polyphylum, then a taxon Metazoa would have to be eliminated from the system of organisms and the name Metazoa would have to disappear.

A correct description of the facts is as follows: a unique origin or a single source of the Metazoa, if they form a monophylum; multiple origin of heterotrophic multicellular organisms, if the Metazoa were a polyphylum. The etymological anomalies "monophyletic origin" and "polyphyletic origin" should, in general, be avoided.

The use of the cited word combinations "molecular phylogeny" and "molecular systematics" is quite common in Anglo Saxon literature. They may give the impression that vital, new disciplines are being established – and perhaps they are intended to do just this. In reality, however, molecular phylogeny and molecular or biochemical systematics cannot exist – any more than it is possible to have a physiological, ethological, or morphological phylogeny or systematics.

The terms phylogeny and systematics refer to entire organisms – not to their parts, such as molecules, mitochondria, cells, or organs. Phylogeny and systematics deal with the genesis of populations of individuals. They try to reveal kinship relations and to represent them in a system by uniting biopopulations of individuals to species and descent communities. All branches of biology are invited to participate in phylogenetic research so that systematics can draw on a variety of different features, from molecules to behavioral patterns. No one, however, can monopolize phylogenetics and systematics, or claim to be the preeminent provider of features.

There is, correspondingly, no justification for a separate and independent discipline of "computer cladistics", but computers can certainly assist in data analyses for the purposes of phylogenetic systematics (cladistics).

The Integration and Meaning of the Fossil Record

If it is our professed goal to discover the kinship relations of all organisms on earth and to adequately reproduce them, there can be only one phylogenetic system for both extant and fossil organisms.

There are, of course, major differences in the value of extant and extinct organisms for the study of kinship relations. Present-day fauna and flora are the basis of our study, because only they can provide a kinship analysis with the necessary spectrum of features. Contrary to popular belief, fossils do not give us the key to the discovery of phylogenetic kinship relations. Extinct species can be systematized only when the fossil exhibits at least one feature which belongs to the autapomorphies of a descent community that has extant species. It is only then that fossilized species can be integrated in the stem lineages of descent communities. Geological records that fulfill this requirement are priceless in two respects:

– The existence of autapomorphies in well dated fossils gives reliable information about the chronological development of evolutionary novelties.

– The oldest fossil with an autapomorphy determines the minimum geological age of the descent community in question. This fossil species establishes the "terminus post quem non" (HENNIG) – the point in time after which the descent community and its sister group could not have originated. We will take a closer look at these three points – the integration of fossils into descent communities, the sequence of evolution of new features, and the age of descent communities.

When dealing with the **integration of fossils** I support the concept of stem lineage (Ax 1984, 1985, 1989b), and will specify the term as a continuation of the preceding arguments (p. 26).

Simple Stem Lineage

Two species B and C as well as the stem species w common to them alone form the smallest possible descent community I (Fig. 4). Since there is only one stem species, its life span or lineage becomes the stem lineage of the community.

We will now place the monotremes *Tachyglossus aculeatus* and *Zaglossus bruijni* in the positions of B and C as sister species (Fig. 10A). This hypothesis implies a stem lineage for the descent community Tachyglossidae in which the spiny coat and claws for cleaning located on the hind-limbs developed as evolutionary novelties [autapomorphies (9) on p. 36 and in Fig. 8]. This stem lineage could actually have been the lineage of a single species, w, but we do not know this. There are no fossils, and even if there were, we cannot say whether they would provide sufficient evidence or not.

It would seem legitimate, in theory at least, to consider a simple stem lineage consisting of one species to be a common phenomenon.

Compound Stem Lineage

On the other hand, we can safely conclude from fossil evidence that there are countless cases of stem lineages consisting of a whole series of stem species.

Let us take the mammals again as an example. The stem lineage of recent mammals stretches from the Upper Carboniferous period to the turn of the Jurassic/Cretaceous period. We have evidence of 250 fossil mammal species which lived during this span of 150 million years. It was during this period that the approximately 60 evolutionary novelties of the Mammalia, a few of which we have already mentioned [autapomorphies (5) on p. 36 and in Fig. 8], were developed. Within the compound stem lineage itself we must differentiate between two groups of species (Fig. 10B).

– Species, whose biopopulations directly succeed each other in a continuous lineage and conti-nually lead from the first stem species, a, of the Mammalia to the last common stem species, h, of recent mammals.
– Species and monophyletic species groups which in the succession of numerous speciations appear to be "lateral" branches of the continuous stem lineage.

It should be more clear that only the evolutionary novelties which developed in the continuous lineage of the species a-h can be taken over in the ground pattern of the recent representatives. Unfortunately, there are no methods available in practice to help us recognize the stem species of

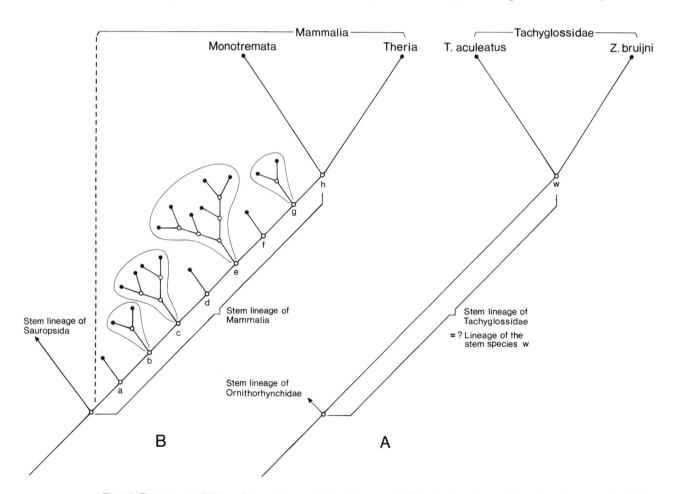

Fig. 10. The two possibilities of the existence of stem lineages. A. Simple stem lineage. The descent community of the Tachyglossidae has only two recent members, the sister species *Tachyglossus aculeatus* and *Zaglossus bruijni*. Theoreti-cally, their stem lineage can be composed of the life span of a single species, namly of the lineage of the stem species w. B. Compound stem lineage. The descent community of the Mammalia consists of the sister groups Monotremata and The-ria. Their stem lineage includes numerous species. In the diagram, only those stem species which succeed each other in a continuous line lead to the recent mammals. The other species and species groups form lateral branches.

a continuous stem lineage. This is only logical, because we cannot expect them to manifest special features not present in recent representatives. On the other hand, species and species groups from lateral branches can, of course, be very easily distinguished with the help of autapomorphies.

The methodical difficulties encountered when determining the position of individual fossils in the stem lineage make me very cautious in my use of terminology. I call all fossils of a descent community which has recent species the "representatives of the stem lineage" or "stem lineage representatives" for short.

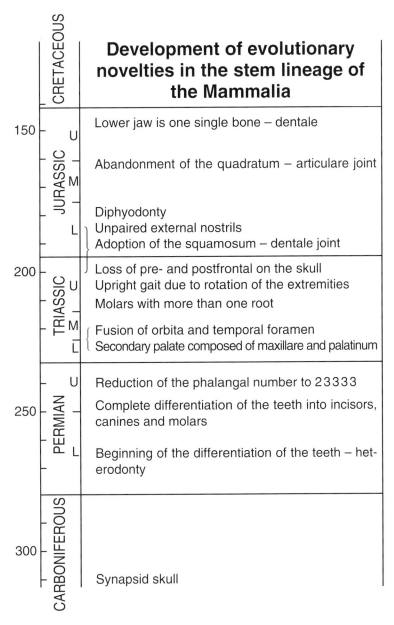

Fig. 11. The successive evolution of fossilizable autapomorphies in the phylogenesis of the Mammalia. (Various authors, in Ax 1988)

We now come to the **sequence in the evolution of new features**. In cases where there is a favorable fossil record, significant stages in the formation of the ground pattern of a descent community can be identified. In this way, numerous fossilized autapomorphies from the stem lineage of the Mammalia which developed in different periods of the earth's history (Fig. 11) are on record. This begins in the Upper Carboniferous period with the synapsid skull, which possesses only one lateral temporal foramen behind the eye socket. The early evolution of heterodonty in the Permian period was followed considerably later in the Triassic period by the upright position, which occurred by means of rotating the extremities under the body. At the end of the stem lineage at the turn of the Jurassic/Cretaceous period, the lower jaw developed into one single bone, the dentale.

This brings us to the **geological age of descent communities**. There is no justification for defining the extent of a descent community according to arbitrarily selected criteria; for example, to say that the Mammalia begin at the border of the Triassic/Jurassic period with the evolution of the secondary jaw articulation between the dentale and squamosum. Nature has determined the extent of each descent community with the splitting of that species which gave rise to its lineage and that of the sister group. We can get close to Nature's decisions by tracking down the oldest fossil which shows at least one autapomorphy of a descent community with recent representatives. This fossil shows the minimum age of that community. The existence of a synapsid skull in the Carboniferous period proves for certain that the descent community of the Mammalia has existed on this earth for over 300 million years, and fossil animals with this skull are stem lineage representatives of the Mammalia – they are mammals.

Let us close by taking a short look at the problem of an optimal order of fossils compared to recent representatives of their descent community. One can easily line up the fossils in a relationship diagram according to the successive increase of autapomorphies they exhibit. A corresponding sequencing is also suggested for the written system (PATTERSON and ROSEN 1977). The fossil species and species groups are simply listed one after the other. Only in the recent representatives of the descent community is subordination used, with consequent indentation of adelphotaxa.

The primary purpose of this textbook is to systematize recent Metazoa – and so I will refrain from more details about the fossils by referring the reader to the literature mentioned at the beginning of this section.

Problematic Large Divisions of Organisms

"Animalia" – "Plantae"

Life is found in animals and plants
ARISTOTLE (BARNES 1985; 2. p.1251)

The first formal classification of organisms into animals and plants is found in the first edition of LINNÉ'S **Systema Naturae** (1735), in which the kingdoms of Animalia and Plantae are established. What is the basis for this seemingly obvious division of organisms? The only consistent difference exists in the two different ways of acquiring the chemical energy essential to life. We call nucleated, green, unicellular and multicellular organisms, that photosynthesize and produce high-energy organic compounds, plants. Colorless unicellular and multicellular organisms which draw their energy from the food they ingest are, with the exception of fungi, animals.

This is the well-known alternative "autotrophy – heterotrophy", which is, of course, useful for a key. However, we need to have precise concepts relating to the plesiomorphy and apomorphy of this alternative in order to interpret plants and animals as possible descent communities of Nature. The anucleate bacteria form the out-group. There are two competing hypotheses about the evolution of chloroplasts – the endosymbiotic hypothesis and the successive hypothesis (see also SCHWEMMLER 1979; MARGULIS 1981; MÖHN 1984; KLEINIG and SITTE 1992; KÄMPFE 1992).

According to the endosymbiotic hypothesis, the colored plastids responsible for photoautotrophic metabolism developed several times independently through a process by which cyanobacteria entered into primary, heterotrophic, nucleated, unicellular organisms. Heterotrophy is thus the plesiomorphy of the alternative and, as such, unsuitable for establishing a phylogenetic unity, Animalia. On the other hand, as a multiply, convergently evolved phenomenon, autotrophy is also not suited to establish green plants as a valid system unity.

The successive hypothesis proposes exactly the opposite process of an early, unique evolution of chloroplasts by way of an intracellular differentiation. Heterotrophy is thus the apomorphy of the alternative, developed obviously more than once. The successive hypothesis rules out autotrophy as a useless plesiomorphy to explain green plants as a descent community. A multiple, independent evolution of heterotrophy for animals is, for obvious reasons, also useless for justifying a monophylum, Animalia.

Whichever way one looks at it, the alternative "autotrophy – heterotrophy" is unsuitable for a basic phylogenetic systematization of nucleated organisms. Animals and plants simply do not exist as descent communities in Nature. As artificial groupings, "Animalia" and "Plantae" have no place in the phylogenetic system of organisms.

"Procaryota" – Eucaryota

A further dichotomy, which emerged comparatively late in biology, is CHATTON'S (1937) division into anucleate Procaryota (bacteria, cyanobacteria) and nucleated Eucaryota (plants, animals). Today, this distinction is regarded as the most fundamental, systematic division of organisms. No textbooks, however, ask the question: to what extent are the Procaryota and Eucaryota two valid system unities, i.e., the equivalents of descent communities in Nature?

In the case of the following alternative features, the manifestations in the Eucaryota are clearly apomorphies.

"Procaryota"	Eucaryota
– Plesiomorphy –	– Apomorphy –
Cell size: a few micrometers	Cell size: from many micrometers to millimeters
—	Compartmentalization: protoplast internally divided by membranes (endoplasmic reticulum)
—	Nucleus with double membrane
A circular DNA molecule free within the cytoplasm. No histones	Chromosomes in the cell nucleus: several linear DNA molecules with histones
Direct cell division	With mitosis. Division of the nucleus before cell division
Parasexuality. Transfer of parts of DNA molecule from donor cells to receptor cells. Recombination through conjugation	Sexual reproduction through the unification of gametes. Genetic recombination through meiosis and syngamy
—	Actomyosin
—	Cytoskeleton made up of protein filaments
—	Mitochondria
—	Chloroplasts
—	Golgi apparatus (dictyosome)
—	Intracellular undulipodium (cilium, flagellum) with microtubules/dynein system. Nine peripheral double tubules and two central single tubules (9 x 2 + 2 pattern). Basal body (centriole) underneath the cell membrane

The mitochondria and chloroplasts, the Golgi apparatus and undulipodia with the 9 x 2 + 2 pattern may have first developed within subgroups of the Eucaryota (LEIPE and HAUSMANN 1993).

The other apomorphies, on the other hand, can surely be interpreted as evolutionary novelties of a stem species common to all eukaryotes. In other words, the Eucaryota are established through numerous autapomorphies as a huge descent community of organisms with nuclei – a monophylum of the system in which autotrophic "plants" and heterotrophic "animals" are unified, regardless of whether they are unicellular or multicellular. On the other hand, no conclusions can be drawn from the plesiomorphies of the prokaryotes. The only feature that could be an apomorphy is the bacteria cell wall made of the peptidoglycan murein which is absent in all eukaryotes. The murein cell wall can not, however, be postulated for the ground pattern of a unity Procaryota, because it is present only in the eubacteria, and not in the archaebacteria. Regardless of this, the absence of murein in the Eucaryota could be an apomorphous state, resulting from a breakdown of the solid murein cell wall as evolutionary developments led to an increase in cell volume.

This means that there is not one single feature which can be interpreted as an autapomorphy of a unity Procaryota. By grouping primary anucleate unicellular organisms under the name "Procaryota", a paraphyletic group of organisms, based exclusively on original features, was created; this group does not form a part of the phylogenetic system.

"Protozoa" – Metazoa

In his classical genealogical tree of organisms, HAECKEL (1866) places a kingdom Protista between the kingdoms Plantae and Animalia. In his Gastraea theory, which followed soon after, Haeckel offered a "phylogenetic classification of the animal kingdom" (1874) in which he took the heterotrophic unicellular organisms out of the Protista and transferred them, as Protozoa (GOLDFUSS 1818), to the animals. The Protozoa was the counterpart of a newly created unity, Metazoa, which comprised all multicellular animals. This arrangement of animals into two "main divisions", "the older, lower group of Protozoa (Urthiere) and the younger, higher group of Metazoa (Darmthiere)", is still found in zoology textbooks today.

Since an animal kingdom based on autapomorphies does not exist, it stands to reason that the artificial product Animalia cannot be phylogenetically subdivided, even if the Protozoa and Metazoa are given attractive labels such as HAECKEL'S "main divisions", or, as in many contemporary textbooks, "subkingdoms".

"Heterotrophic unicellular organisms" are not a unity of the phylogenetic system. Unicellularity is the plesiomorphy in the alternative to multicellularity in the Metazoa. "Protozoa" are a paraphyletic collection of primarily unicellular, heterotrophic Eucaryota.

We should not, however, throw the baby out with the bath water. Several subgroups of the paraphylum Protozoa can be indisputably established as monophyletic unities.

Among the traditional Sporozoa, the Telosporidia (Apicomplexa) are characterized by a series of autapomorphous features. These include the haplohomophasic alternation of generations with gamogony and sporogony, the formation of a solid spore through the zygote, and the formation of a unique apical complex (penetration organelle) in infection germs.

In the ground pattern of the monophylum Ciliata, the following characteristics are regarded as autapomorphies: the complete ciliation of the cell, conjugation with the fusion of haploid migratory and stationary nuclei, and the division of the diploid synkaryon in micronuclei and macronuclei.

For the rest, the phylogenetic systematization of heterotrophic, unicellular organisms is still only in its infancy. No real benefit can be expected from the "revised classification of the Protozoa" (LEVINE et al. 1980) undertaken by a committee of 16 experts, as long as classification is based on "convenience" (which cannot be objective), and "does not necessarily indicate evolutionary relationships".

On the other hand, the monophyly of a unity, Metazoa, can be excellently justified. All multicellular, heterotrophic animals can be traced back to one stem species which is common only to them (p. 54).

"Invertebrata" – Vertebrata

The alternative "Protozoa" – Metazoa is not the only choice scientists have developed over the years. Man's apparent need to express the order in Nature in terms of dichotomous divisions has also bestowed zoology with the division of animals into those with and those without a vertebral column.

The combination of certain heterotrophic unicellular and multicellular organisms into Invertebrata is based on the absence of a feature, namely of the vertebral column, and thus of the constitutive element of all those animals for which LAMARCK (1794) created the name Vertebrata. The absence of a vertebral column is, however, undoubtedly an original phenomenon. In other words, the consolidation of all animals without a vertebral column in the Invertebrata created a new paraphylum. The artificial character of this division can be easily seen by looking at the arbitrary cut the group Invertebrata makes through the monophylum Chordata with the subtaxa Tunicata, Acrania and Vertebrata (Craniota). The tunicates and lancelets are animals without vertebral columns and, for this reason, feature in textbooks on the invertebrates. They are, however, more closely related to the Vertebrata than with any other random group of the "Invertebrata".

In this context, we must, again, insist on a clear position even if, unlike the Protozoa, the Invertebrata are not considered to form an explicit system unity. The fact of the matter is that this division is commonplace in the literature; HYMAN'S classical multivolumed work bears the title THE INVERTEBRATES, as do numerous other, more modern English textbooks. Similarly, in the German literature all editions of the first volume of KAESTNER'S TEXTBOOK OF SPECIAL ZOOLOGY (LEHRBUCH DER SPEZIELLEN ZOOLOGIE) are entitled INVERTEBRATES or INVERTEBRATE ANIMALS (WIRBELLOSE or WIRBELLOSE TIERE, 1954–1993).

The Inflation of Kingdoms

"How many are the kingdoms of organisms?" (LEEDALE 1974)

HAECKEL'S division of organisms into three groups, Plantae, Protista and Animalia, was for a long time the most elaborate grouping of organisms into kingdoms. This changed only around the middle of this century. COPELAND (1938) distinguishes four kingdoms, WHITTAKER (1959, 1969) five kingdoms, and LEEDALE (1974) derives 18 kingdoms of eukaryotic organisms from the kingdom of Procaryota (Monera).

WHITTAKER'S FIVE KINGDOM SYSTEM divides organisms into the kingdoms Procaryota, Protoctista, Plantae, Animalia and Fungi. This division has become particularly popular thanks to MARGULIS and SCHWARTZ (1988, 1989), so that a closer analysis of its rationale would seem to be especially worthwhile.

To begin with, two kingdoms are clearly paraphyletic groupings, the "Procaryota" and the "Protoctista". The latter "comprise the eukaryotic microorganisms and their immediate descendants" (MARGULIS et al. 1988, p.77). The problem of the monophyly of system unities is apparently of no great relevance to the other three kingdoms either. In "A phylogeny of life based on the Whittaker five-kingdom system..." (MARGULIS et al. 1988, p.II), the Plantae and Fungi trace back to the Protoctista, each in two lineages, and the kingdom Animalia originates even in three separate lineages. The presentation thus logically implies the "polyphyly" of the three kingdoms Plantae, Animalia and Fungi.

This, however, is only one side of the coin: the non-fulfillment of the requirements necessary for setting up valid system unities. In order to attain theoretical clarity, we must flip the coin and assume that three, four, five, or indeed any number of kingdoms into which organisms have been divided can really, kingdom for kingdom, be established as monophyla. It is not, however, reconcilable with the hierarchical order in Nature ensuing from phylogenesis to offer each of these "kingdoms of organisms" an identical rank.

One can hypothesize that all cellular organisms form one huge descent community based on elementary similarities (DNA, proteins). Only sister groups that originated at the same time can have the same rank in this unity. In other words, only two of the many organism kingdoms could be categorized as Regnum. We, however, go as far as to reject this, too. Just as all categories are unusable (p. 20), so, too, is the category Regnum irrelevant to the phylogenetic system.

The question of the number of organism kingdoms is not a problem for systematic biology. Kingdoms exist in the human mind, not in Nature.

The discussion about organism kingdoms has yet another facet relevant to the central theme of this book. In the classifications drawn during this century, the term "animal kingdom" refers basically to multicellular organisms, sometimes including, sometimes excluding the Porifera. In the former case, the constituents of the "animal kingdom" are identical to the constituents of the monophylum Metazoa – the names Animalia and Metazoa become synonymous. One could use the older name Animalia and limit it to the multicellular organisms. This might, of course, lead to further misunderstandings, so that it is probably sensible to continue using the name Metazoa for that monophyletic unity for which HAECKEL created it more than 120 years ago.

Systematization of the Metazoa

Overview

We will place the highest-ranking sister group relations of the Metazoa at the beginning in a relationship diagram (Fig. 12), and in the following hierarchical tabulation.

> **Metazoa**
> **Porifera**
> **Epitheliozoa**
> **Placozoa**
> **Eumetazoa**
> **Cnidaria**
> **Acrosomata**
> **Ctenophora**
> **Bilateria**
> **Spiralia**
> **Plathelminthomorpha**
> **Euspiralia**
> **Radialia**
> **"Tentaculata"**
> **Deuterostomia**

These two representations show phylogenetic kinship unmistakably without relying on category labels. Categorization of the Porifera, the Placozoa, the Cnidaria, and the Plathelminthes (or whatever) as phyla with identical ranking is not relevant here. The very fact that they occupy entirely different positions in the hierarchy of the phylogenetic system makes it necessary to repeat once again that categories are irrelevant to this system.

Our main concern is to meet two requirements:
– Every name in the phylogenetic system represents a descent community in Nature or its equivalent monophylum in the system. Every named unity must be justified by listing autapomorphies that are evolutionary novelties particular only to that unity. Where this proves to be impossible for the present, the names, of those taxa whose provisional inclusion seems purposeful will be put in quotation marks, as illustrated by the "Tentaculata" in our overview.
– Sister group relations will be indicated in relationship diagrams by grouping two names on the same level, and in the hierarchical tabulation by equal indentation of the two names. Reasons for each hypothesized adelphotaxa relation must be presented.

These two requirements determine the following representation. There is a constant switching from the characterization of individual monophyla to the ensuing representation of their highest-ranking adelphotaxa relation. Hence, the justification of the Metazoa as a monophylum is followed by a discussion of the adelphotaxa relation Porifera – Epitheliozoa, and the justification of the monophyly of the Epitheliozoa is followed by the representation of the sister group relation Placozoa – Eumetazoa, etc. Orientation is ensured at each point in the system by the interlinked relationship diagrams.

The autapomorphies of the individual monophyla are each combined in a black block; they stand in the stem lineages in which they developed as evolutionary novelties. The block numbers are parenthesized in the text below the captions of the corresponding autapomorphies.

The first attempt at writing a consistent phylogenetic system of the Metazoa does not yet offer a sufficiently balanced treatment of the individual taxa. For example, further systematization of the Anthozoa, Hydrozoa and Scyphozoa as subtaxa of the Cnidaria is not yet possible because no corresponding research has yet been published in the literature of phylogenetic systematics. This situation ought to be changed as quickly as possible. In contrast, the Plathelminthomorpha can already be represented with the Gnathostomulida and the Plathelminthes in exemplary manner, promising a better future for textbooks on the phylogenetic system of the Metazoa.

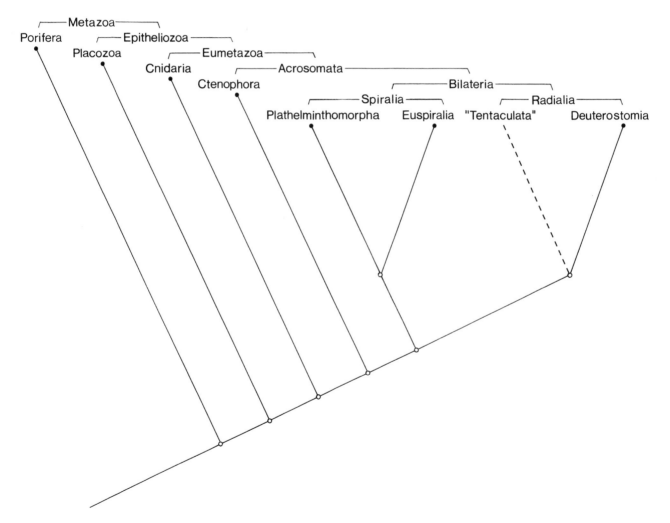

Fig. 12. The highest-ranking adelphotaxa relations in a diagram of phylogenetic kinship of the Metazoa. All taxa mentioned are justified on the basis of evidence that relies exclusively on their own autapomorphies (exception "Tentaculata").

Monophyly and Ground Pattern

When justifying the monophyly of system unities and working out their ground patterns, we usually, more or less immediately, draw upon the subordinated adelphotaxa. The reason for this is that we need the synapomorphies between them in order to arrive at the autapomorphies of the superordinated unity in question. Since we cannot deal with everything at once, the adelphotaxa themselves appear only later in the representation. This explains why the subordinated adelphotaxa Porifera and Epitheliozoa are adressed earlier while justifying the Metazoa as a monophylum, whereas they themselves are dealt with only several pages later.

■ **Autapomorphies**
(Fig. 18 → 1)[2]

(Original components, which have been adopted from the level of unicellular organisms into the composition of certain unique features of the Metazoa, are placed in parentheses)

= Synapomorphies of the adelphotaxa Porifera and Epitheliozoa.

– Multicellular, gonochoric (aquatic) organism with diploid body cells.

– Gametic meiosis with the production of haploid egg and sperm cells.

– Oogenesis: one fertilizable egg cell and three abortive, plasma-deficient polar bodies develop from one oocyte (Fig. 13).

– Spermatogenesis: four structurally identical sperm develop from one spermatocyte.

– Sperm structure (Fig. 14)
1. Head with nucleus and several small acrosome vesicles (EHLERS 1993).
2. Middle piece with four large mitochondria and two centrioles arranged at right angles. The distal centriole forms the basal body of the cilium.
3. Terminal filament with one cilium (the cilium has the original 9 x 2 + 2 pattern of the Eucaryota with two central microtubules and nine peripheral double tubules).

– Unification of male and female gametes into one diploid zygote. The gametes are freely released; they inseminate the egg cells in water.

– Impermeable cell-cell connections (= occluding contacts): septate junctions in the "invertebrates"; tight junctions in the vertebrates (BARTOLOMAEUS 1993a).

– Production of an extracellular matrix (ECM) with collagen fibrils.

– Ontogenesis with radial cleavage during the arrangement of the blastomeres to the blastula.

---

2 Throughout the book the number following the arrow refers to the stem lineage in which the autapomorphies originated as evolutionary novelties.

- Adult a flat, crawling blastaea (placula). The monophasic life cycle without larvae is a plesiomorphy.

- Somatic differentiation. Minimally present:
  1. Outer ciliated cells for locomotion and food intake (?osmotrophy). (As an original feature, the equipment with one cilium and one striated ciliary rootlet was adopted from the level of unicellular organisms; cf. p. 59).
  2. Inner aciliated stem cells (archaeocytes, neoblasts).

These features will be hypothesized point by point as evolutionary novelties in the ground pattern of the Metazoa. They evolved in the stem lineage of a descent community in Nature which we call Metazoa. Each of the evolutionary novelties was realized in the feature pattern of a concrete organism which lived in the early Precambrian period as the last common stem species of all recent Metazoa.

A convincing justification of the monophyly of the Metazoa is derived from the complex of features relating to egg and sperm formation. When comparing the gametic meiosis, the severing of tiny polar bodies during oogenesis, and the differentiation of a certain sperm pattern, which correlates to the primary free release of sperm, there are detailed congruences between the Porifera and diverse subgroups of the Epitheliozoa. The most parsimonious explanation is the assumption of a unique evolution in a stem lineage common to all Metazoa. A justifiable alternative does not exist. There is an equally convincing argument for the hypothesis of the single evolution of an extracellular matrix (ECM) and of the collagen fibrils found in this matrix.

Insufficient knowledge of *Trichoplax adhaerens* (Placozoa) leads to difficulties. Only fragmentary observations of sexual reproduction with oogenesis and spermiogenesis are available (p. 80); ECM and collagen fibers are absent. As long as the life cycle of *Trichoplax adhaerens* has not been fully elaborated, it would seem pointless, however, to operate with negative assertions of questionable value.

In the process of ontogenesis, the radial cleavage and development of the germ up to the blastula stage are identical in numerous species of Porifera and Epitheliozoa. Both phenomena are taken as autapomorphies in the ground pattern of the Metazoa. It should, however, be noted that a justification using the out-group comparison with unicellular organisms is not possible. The same holds true for the manifestation of totipotent aciliated cells (archaeocytes of the Porifera, neoblasts of the Epitheliozoa).

Further comparisons of the stages of ontogenesis with diverging somatic differentiation, or of certain manifestations in the adults cannot, on the other hand, be made. Let us take the nutritional system as an example. The filter apparatus of the Porifera, which has collar cells (choanocytes), is completely different from the intestinal epithelium of the Epitheliozoa, which has glandular cells producing digestive enzymes. There is no basis for homologization. Consequently, neither the one nor the other organization can be put in the ground pattern of the Metazoa.

The stem species of the Metazoa was presumably a gonochoric organism. The evolutionary change from gonochorism to hermaphroditism can be seen in numerous taxa, and can be explained using different models (GHISELIN 1969, 1974b). A development of gonochorists from

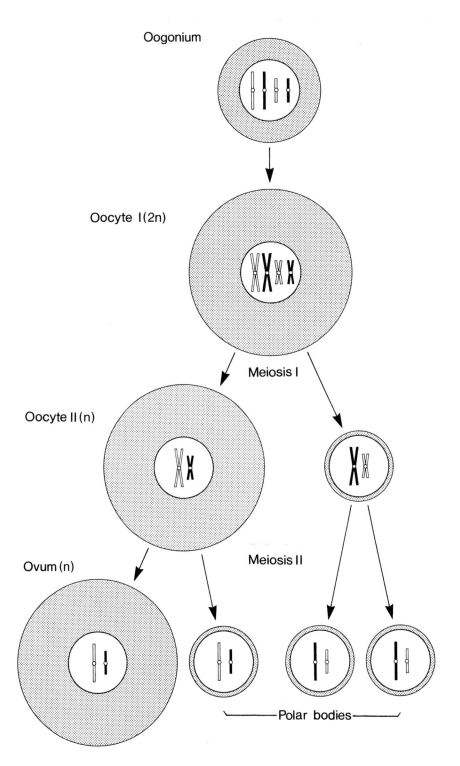

Oogonium

Oocyte I(2n)

Meiosis I

Oocyte II(n)

Meiosis II

Ovum(n)

Polar bodies

Fig. 13. Oogenesis in the ground pattern of the Metazoa. Diploid oogonia begin the meiotic phase during the storage of reserve substances. One oogonium divides twice, producing one haploid egg rich in nutrients, and three haploid polar bodies low in plasma.

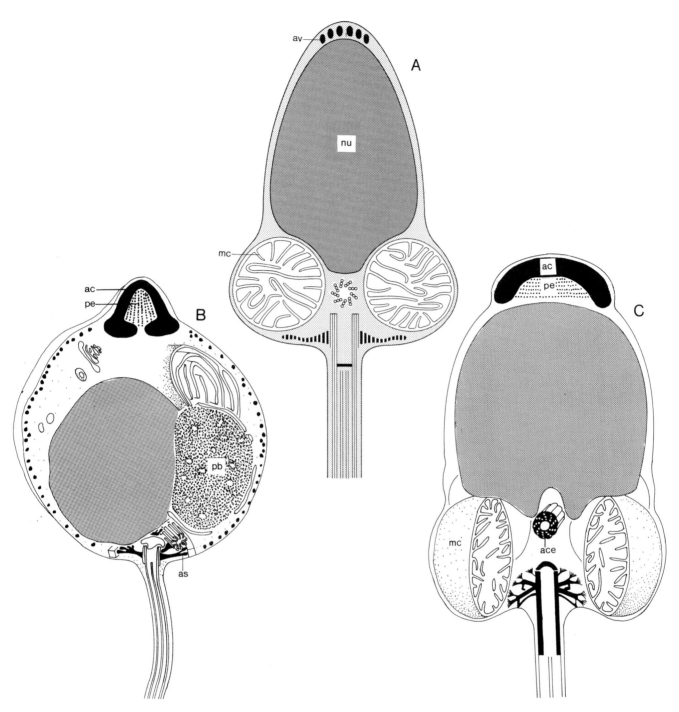

Fig. 14. Comparison of the sperm structure in the ground pattern of the Metazoa (A) with features in the Acrosomata (B and C). A. *Halammohydra schulzei* (Cnidaria, Hydrozoa). Sperm with several acrosomal vesicles in the head in front of the nucleus, four large mitochondria behind the nucleus, a terminal cilium, and an accessory centriole in front of the basal body. B. *Beroe ovata* (Ctenophora), and C. *Priapulus caudatus* (Priapulida) represent a commonly found state in the Acrosomata with external fertilization in ocean water. In both cases, the acrosomal complex consists of a uniform, convex acrosome and an adjacent perforatorium, which is primarily formed from granular material. An acrosomal complex of this nature can be interpreted as an autapomorphy of a unity Acrosomata. ac = acrosome. ace = accessory centriole. as = anchor structure. av = acrosomal vesicle. mc = mitochondrium. nu = nucleus. pb = paranuclear body. pe = perforatorium. (A Ehlers 1993; B Franc, in Hernandez-Nicaise 1991; C Afzelius and Ferraguti 1978. B and C completed after Bacetti 1979)

hermaphrodites is, on the other hand, provable only in individual cases, such as in the gonochoric trematode *Schistosoma haematobium* and related species within the hermaphroditic Plathelminthes.

The stem species of the Metazoa had a monophasic life cycle without larvae. It is very probable that the adult was a benthic organism in the form of a flattened blastaea without axes of symmetry (placula BÜTSCHLI 1910). This argumentation draws on the comparison of the primary biotope ties of species of high ranking subgroups of the Metazoa – the Porifera, Placozoa, Cnidaria, Ctenophora and Bilateria. *Trichoplax adhaerens*, as representative of the Placozoa, is a flat, diploblastic organism without larvae which crawls on the substrate. Correspondingly, we can postulate a freely moving, bilaterally symmetrical organism in the benthic zone, and the primary absence of larvae for the stem species of the Bilateria. This is possible simply because the life-form in question here is realized in many primarily simple Bilateria such as the Gnathostomulida, Plathelminthes, Nemertini, or the Gastrotricha. The Porifera and the Cnidaria each have a biphasic life cycle with a sessile adult and a freely swimming plankton larva. In this comparison, the monophasic Ctenophora form the only holoplanktic unity; the secondary change of some Ctenophora into the benthic zone (*Coeloplana, Ctenoplana*) is not relevant in this context.

The hypothesis that the Metazoa have a benthic stem species is based on our knowledge of the Placozoa, and the interpretation of the stem species of the Bilateria as a monophasic, vagile organism in the benthal. It is difficult to conceive of a round blastaea inhabiting soil. Based on the facts, we hypothesize a disk-shaped organization in the ground pattern of the Metazoa, as is realized in *Trichoplax adhaerens*.

The life-forms of the Porifera, Cnidaria, and Ctenophora then evolved independently of each other out of vagile, soil-dwelling organisms. The biphasic life cycles of the Porifera and Cnidaria developed twice independently. In the stem lineage of the Ctenophora, a separate way led from the bottom into open waters, to a purely planktonic organism.

The Search for the Adelphotaxon
Among Unicellular Organisms

No recent species and no recent species group can be the phylogenetic ancestor of some other recent species or unity of species. Commonly held views as to the origin of the Metazoa from certain currently living, unicellular organisms are incompatible with our knowledge of the process of phylogenesis. This holds true for the attempt to derive the Metazoa from the Choanoflagellata (Craspedomonadina) through the Porifera, as well as for the assumption that they originated from the Ciliata through the plathelminth unity of the Acoela.

The Metazoa do, however, have an adelphotaxon among the Eucaryota living today. We must try to establish a recent species or species group as the closest phylogenetic relative of the Metazoa, provided, of course, that the resulting hypothesis can be justified.

The search for the adelphotaxon of the Metazoa can be limited to the following, general aspects:
– Our principle of parsimony dictates that the search for the adelphotaxa should be among heterotrophic, unicellular organisms.

– Tissues made up of cells with one cilium are widespread among the Metazoa[3] (Porifera, Placozoa, Cnidaria, Gnathostomulida, Brachiopoda, Pterobranchia, Echinodermata etc.). This monociliarity is indisputably interpretable as a plesiomorphous state within the Metazoa. Multiciliarity with several cilia per cell, such as in the Plathelminthes or the Nemertini, forms the apomorphous alternative. Thus, the adelphotaxon of the Metazoa should be found among unicellular organisms with a single cilium.

– In the ground pattern of the Metazoa, a basal body and a striated ciliary rootlet in the interior of the cell belong to the cilium with the 9 x 2 + 2 microtubule pattern; an accessory centriole is located at right angles to the basal body (p. 55). Since striated rootlets and the accessory centriole are common among unicellular organisms (PITELKA 1974; WOLFE 1972; VICKERMANN 1991), one can assume that a cilium with these substructures passed on from the unicellular eukaryotes to the stem species of the Metazoa.

Bearing these premises in mind, let us examine the widely held view that a phylogenetic relationship exists between the Choanoflagellata and the Metazoa. This view is caused by the conspicuous similarities between the choanoflagellate cells and the choanocytes inside the body of the Porifera. A collar of microvilli surrounds the single cilium of the cell in both unities; the cilium itself carries wing-shaped projections (vanes) and does not have rootlets in either case. The suspicion of a homology of the structures in the Choanoflagellata and Porifera is, however, relativized by the differences in the ultrastructure. The extremely delicate vanes of the Choanoflagellata cannot be detected in electron microscopic cross sections through the collar, which is very much in contrast to the corresponding pictures of the thick, massive wings of sponges (HIBBERD 1975; MEHL and REISWIG 1991).

The following fact is even more significant. The collarless, monociliated cells of the amphiblastula larva of the Calcarea (calcareous sponges) have a ciliary rootlet with distinctive striations (GALISSIAN and VACELET 1992), and this is also seen in the Placozoa and many unities of the Eumetazoa. "Normal" monociliated cells with a striated ciliary rootlet and without microvilli collars can thus be clearly justified as a feature of the stem species of the Metazoa. This stem species cannot have been an organism composed of rootless choanocytes, since one logical consequence would be the untenable hypothesis of an independent evolution of striated ciliary rootlets within the Porifera and in the other Metazoa.

The facts can only be interpreted to reach the reverse conclusion. The choanocytes of the Porifera evolved as evolutionary novelties in the stem lineage of sponges – with the development of a microvilli collar and the ciliary vanes as well as the degeneration of the original ciliary rootlet. Collar cells are an autapomorphy of the Porifera. They developed independently of similar structures such as are found in the Choanoflagellata and here and there within the Eumetazoa.

The question regarding the sister group of the Metazoa among the monociliated, unicellular, heterotrophic organisms has not yet been answered.

---

3 For the sake of using uniform terminology, we do not make the traditional differentiation between flagellum and cilium, which has become largely invalid in view of the fact that both share the same ultrastructure. The terms "monociliated cells–multiciliated cells" clearly comprise the alternatives we wish to describe here.

Mesozoa

– Parasitic Metazoa with Undetermined Kinship –

Before going on to explain the first, basal dichotomy of the Metazoa in the Porifera and Epitheliozoa, I would like to look at the Mesozoa. The Orthonectida and the Dicyemida, two taxa comprising simply organized parasites of marine invertebrates, are conventionally united under this name. They have not, as yet, found a secure position in the system of the Metazoa.

■ **Autapomorphies**

– Body is made up of a cellular, ciliated epidermis and a cellular interior without digestive tissue in which the sexual cells are produced.

– Endoparasites with intake of dissolved organic substances through the epidermis.

– Cilia of multiciliated epidermal cells each with one rootlet lying horizontally under the cell surface. The tip of the ciliary rootlet is pointed towards the front.

Orthonectida

■ **Autapomorphies**

– Epidermal cells arranged in rings.

– Epidermis only partially ciliated.
  Rings with wholly or partially ciliated cells alternate with rings of aciliated cells.

About 20 species are described which appear as tissue parasites in a wide spectrum of marine host organisms of the taxa Plathelminthes, Nemertini, Annelida, Gastropoda, Bivalvia, Ophiuroida and Tunicata.
With only one exception, the Orthonectida are gonochoric. The female is between 200 and 250 μm long; the males are distinctly smaller. Only *Stoecharthrum giardi* from the polychaete *Scoloplos armiger* is a hermaphrodite. This is a further example in which the change within a monophyletic taxon from gonochorism to hermaphroditism can be clearly proven.
The epidermis of sexually mature specimens consists of multiciliated cells and aciliated cells which cover the body in rings; the number and arrangement of the ciliated and aciliated rings are constant in the individual species. The epidermis surrounds an inner cavity which is filled with cells. The inner cavity of the females mainly contain egg cells. Muscle cells are located in the periphery. The male has one testis. Aside from undifferentiated cells, the male also features muscle cells.
The most analyzed species, *Rhopalura ophiocomae* and *Ciliocincta sabellariae* (Kozloff 1965, 1969, 1971), will be treated in greater detail in the following passages.

Rhopalura ophiocomae (Fig. 15A-C) lives as a parasite in *Amphipholis squamata* (Ophiuroida). The parasite exists mainly in the area of the bursae near the gonads of the brittle star. The females appear in two sizes, about 250 µm (65-80 µm in diameter) and about 150 µm long. The males are about 130 µm long.

The epidermal cells are arranged in 35 or 36 rings in the larger females; the narrow rings are made up of aciliated cells. A genital pore is located in ring 19; it is surrounded by a wreath of tiny cells. The mass of the axial cells is made up of closely packed oocytes. Between the epidermis and egg cells lies a loose net of longitudinal muscle cells, circular muscle cells, and slanting cells with myofilaments. The longitudinal muscles are located interior to the other muscle cells.

In the male, the epidermis is made up of 20 cell rings. The cells of rings 1, 2, 11, 15, 19, and 20 are completely ciliated, the cells of rings 3, 5, 7, 8, 9 only have wreaths of cilia, and the remaining narrow rings are aciliated, as in the female. Refractive, "crystalline" inclusions are located in the cells of rings 4-9; their composition and function are unknown.

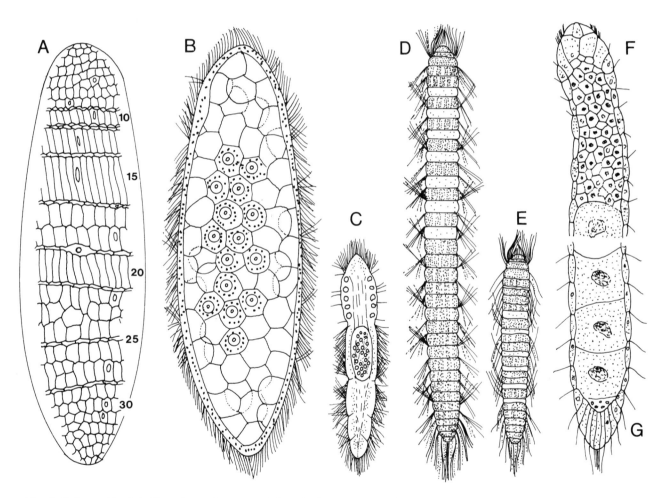

Fig. 15. Orthonectida. A-C. *Rhopalura ophiocomae*. A. Female. Representation of cell boundaries in the epidermis (silver nitrate impregnation). Numbers refer to individual cell rings. Genital pore in ring 19. B. Female with oocytes based on observation of the living organism. C. Male with testes in the middle of the body based on observation of the living organism.
D-G. *Ciliocincta sabellariae*. D. Female. E. Male based on observation of the living organism. F. Female. Anterior end with undifferentiated cells in front of an oocyte. G. Female. Posterior end with a row of oocytes. (Kozloff 1965, 1969)

An unpaired testis lies in the middle section of the body; it is surrounded by its own muscle sheath, which presumably serves to squeeze out sperm. The male genital pore is located at the end of the testes between rings 12-14. The sperm have a round head and a cilium.

Longitudinal muscle cells stretch peripherally along the whole body. Behind the testis, eight longitudinal muscle cells surround a central strand made up of four cells with cross-striated filaments of an unknown nature (perhaps paramyosin).

In the ultrastructure, the organization of the cilia is of particular interest to us. A single, striated ciliary rootlet is located at the basal body of the individual cilia. The rootlet is horizontally positioned and runs closely under the cell surface straight or diagonally towards the front. This is the case in both males and females.

Mature females and males leave the host organism. When they are in open waters, they place their genital openings against one another. Sperm is transferred to the female; the oocytes are internally fertilized. The penetration of sperm into an egg cell causes the release of two polar bodies. The fertilized egg cell develops within the female into a ciliated larva made up of a few cells. The male dies.

Not enough is known about the further course of the life cycle. Even the path of invasion of the larva into a new host is unknown. The multinucleate "plasmodium", which is presumed to originate from a larva or from certain larval cells, seemingly does not belong to the Orthonectida. In fact, it is more likely to be degenerated host tissue (KOZLOFF 1992). The development of adults from larvae in this tissue is unresolved.

We will compare *Rhopalura ophiocomae* with *Ciliocincta sabellariae* (Fig.15D-G), a parasite in the skin of the polychaete *Sabellaria cementarium* (Annelida). *Ciliocincta sabellariae* is suitable for comparative purposes due to substantial differences in terms of habitus, ciliation, and the arrangement of egg cells.

Ciliocincta sabellariae is a rod-shaped, elongated organism. The females reach a length of 265 μm (only 22-24 μm in diameter); the males grow up to 135 μm long (about 20 μm in diameter). The distinctive ringing of the body is due to the predominantly equal-sized rings made up of epidermal cells – altogether 38-39 in the female, and about 20 in the male. The majority of epidermal cells have obvious, granular inclusions; in both sexes, seven cell rings do not have these inclusions.

In comparison to *Rhopalura ophiocomae*, there are no completely ciliated epidermal cells in *Ciliocincta sabellariae*. Cilia are found mainly on the front and/or rear edges of the grana-carrying cells and thus surround the body in the form of cilia wreaths. Rostral ciliary rootlets, as seen in *R. ophiocomae*, are present, and longitudinal muscles and, in some areas, ring muscles are also found. Due to the rod-shaped body, the oocytes are arranged in one long row in the female body. In the female, the first fifth of the axial mass is made up of small undifferentiated cells, while in the male they constitute the first half of the body; the testis here is found in the second half.

Dicyemida

The Dicyemida inhabit the kidneys and the pericardial coelom of Cephalopoda. Of the 75 known species, the findings dealt with in the following relate to the taxa *Dicyema* and *Dicyemennea*. In cases of particularly dense population, the kidney sacs of squid can be filled with hundreds of thin, string-like individuals with an average length of 1-2 mm.

The epidermis is made up of 40-50 completely ciliated cells. They enclose an area which is occupied by a single, long cell – the axial cell. Other cells, called axoblasts, are located inside the axial cell. The axoblasts give rise to embryos and larvae.

Two rows of epidermal cells with short cilia form the calotte on the front end of the body. With these cilia, the parasite anchors itself in the kidney of the squid between the microvilli of the renal epithelium. The rest of the body with its long cilia floats freely in the urine of the host where it draws dissolved, organic substances.

The Dicyemida have two different modes of reproduction, and two technical terms nematogen and rhombogen are used to denote them. In the **nematogen** (Fig. 16A), asexually, wormlike larvae develop from axoblasts, whereas in the **rhombogen**, a hermaphroditic gonad develops. From the fertilized egg cell, a cilia-like larva develops – the socalled **infusoriform**.

How are the nematogens and the rhombogens related? The first individual to take up residence in the kidney of a young squid is a nematogen. Asexually, daughter nematogens develop from it, which, in turn, give rise to further daughter nematogens, leading to the density of individuals mentioned above. A change in the reproductive pattern occurs when the squid reaches sexual maturity. Nematogens change into rhombogens and the worm-like larvae now begin to develop into rhombogens directly.

Cell (axoblast) reproduction within other cells (axial cells) was interpreted as an autapomorphy of the Dicyemida. Let us examine a few stages of this unique process, which cannot be compared with that in any other animal.

The first division of the axoblast and further mitoses within the **nematogen** result in the formation of a cluster of cells in which the future epidermal cells surround a central cell (Fig. 16C-H). This central cell divides to form the axial cell and what will later become the first axoblasts of the new organism. The axoblast is then taken up into the axial cell and divides intracellularly to form more axoblasts. The worm-shaped larva develops by repeated mitotic division and finally leaves the mother organism and becomes a daughter nematogen.

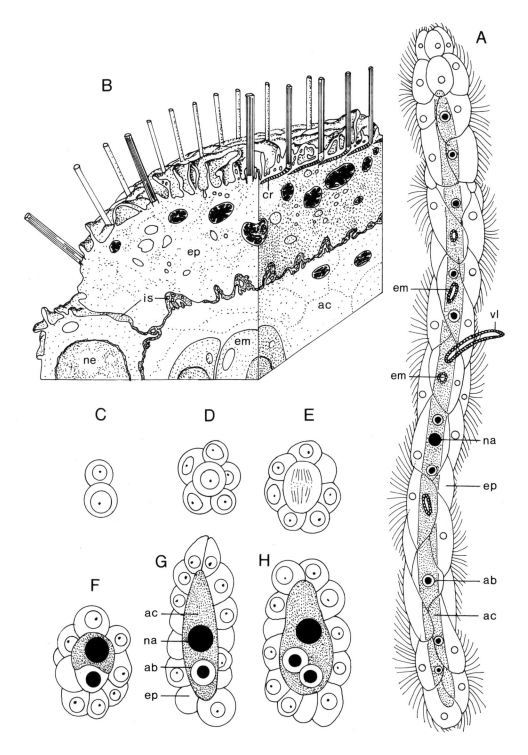

Fig. 16. Dicyemida. A. Nematogen with asexual formation of worm-shaped larvae. B. *Dicyema truncatum*. Ultrastructure of the epidermis and axial cell. C-H. *Dicyema balamuthi*. Early stages of asexual formation of a new nematogen. C. First division. D. Prospective epidermis cells surround a central cell. E. and F. Division of the central cell with formation of the axial cell and the first axoblast. G. First axoblast is taken up by the axial cell. H. Division of the axoblast. ab = axoblast. ac = axial cell. cr = ciliary rootlet. em = embryo in the axial cell. ep = epidermal cell. is = intercellular space. na = nucleus of the axial cell. ne = nucleus of the epidermal cell. vl = vermiform larva. (A Lapan and Morowitz 1972; B Bresciani and Fenchel 1965; C-H McConnaughey 1951, in Kozloff 1990)

The development of the axoblasts in the **rhombogen** takes a different course (Fig.17A-E). The first axoblast division is unequal, and results in two cells of which the smaller cell collapses. The following two divisions produce two further cells. One of these penetrates the parental cell, and produces a small number of aciliated sperm by means of meiosis. The other cell divides to produce several oocytes, which surround the parental cell at the periphery. Insemination takes place in the hermaphroditic gonad. After sperm penetration, meiosis occurs in the oocyte which thereby releases polar bodies.

There are varying morphological interpretations of the accumulation of egg and sperm cells in the rhombogen. The description offered above assumes the presence of hermaphroditic gonads; another interpretation suggests, however, that these are independent individuals.

The **infusoriform** develops from the fertilized egg cell – a larva 40-50 μm long (Fig.17F). Two large cells of the aciliated front end contain light-refracting inclusions (magnesium inositol hexaphosphate). The second half of the body is ciliated. It has a hollow, empty cavity, referred to as the urn.

A rupture of the epidermis releases the larvae, which are then excreted with the urine from the squid. We do not know what happens to them after that. The infusoriform is probably the infection

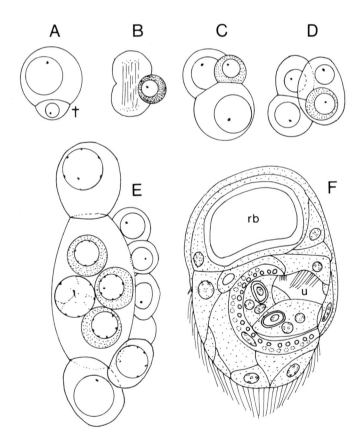

Fig. 17. Dicyemida. A-E. *Dicyema.* Development of the hermaphroditic gonad. A. First division; smaller cell degenerates. B. Second division completed, third division begins. C. Third division completed. D. The smaller cell of the second division has entered the parental cell; the other cell has divided. E. Hermaphroditic gonad. Cells inside the largest cell become sperm (dotted); cells outside form oocytes. F. *Dicyemennea.* Infusoria-like larva. Sagittal section. rb = refringent body. u = urn. (A-E McConnaughey 1951, in Kozloff 1990; F Matsubara and Dudley 1976, in Kozloff 1990)

germ of the Dicyemida. However, no observations are available to account for the development of nematogens from ciliated larvae, or for the primary colonization of a squid kidney by a nematogen.

In comparison with the Orthonectida, the number and orientation of the cilia rootlets are important (Fig. 16B). Just as in the orthonectidans, the ciliated cells of all stages of the Dicyemida have only one ciliary rootlet per cilium. The rootlet lies close under the cell surface with its tip pointed forward (RIDLEY 1968, 1969; MATSUBARA and DUDLEY 1976).

Phylogenetic Kinship

The problem of the kinship relations of the Mesozoa should be discussed in terms of three questions. Do the Orthonectida and Dicyemida belong together? Are they representatives of the monophylum Metazoa at all? If so, where do they stand within the Metazoa?

1. Do the Orthonectida and Dicyemida form a monophylum, or did they – as is often assumed – evolve convergently into similar parasites?
I have already introduced the probable autapomorphies of a monophylum Mesozoa. Orthonectida and Dicyemida are similar in that they have an epidermis made up of multiciliated cells and an interior cavity which does not have digestive tissue and in which gametes are formed and larvae develop. This construction would seem to be unique among animals. The main focus is on the common apomorphous absence of a nutritive endoderm. This is a result of the endoparasitic nature of the organism, which absorbs food in the form of dissolved substances through its body surface. The most likely hypothesis of a single evolution of this type of parasitism and of the body structure related to it complies with the principle of parsimony. This interpretation is not only incontestable, it is also supported by the number and arrangement of the epidermal ciliary rootlets. Each of the cilia in the Orthonectida and the Dicyemida has only one rootlet at the basal body. In both cases, the striated rootlets lie horizontally under the surface of the cell with their tips pointing forward.
These facts speak in favor of a parasitic stem species common to the Orthonectida and the Dicyemida, and hence the monophyly of a taxon Mesozoa.

2. The second issue to be resolved is also controversial. Are the Mesozoa a subtaxon of the monophylum Metazoa, or did they develop independently from heterotrophic unicellular organisms?
Two autapomorphies of the Metazoa, found in the Mesozoa, offer a clear answer. The first is the mode of oogenesis with the release of small polar bodies during meiosis, and the second is the development of septate (tight) junctions between epidermal cells. These features clearly support the proposition that the Mesozoa belong to the Metazoa, and speak against a separate development from unicellular organisms.

3. These answers bring us to the question of the position of the Mesozoa within the monophylum Metazoa.
It is now time to take a more precise look at the multiciliated epidermal cells in the Orthonectida and the Dicyemida. Zonulae adherentes are an essential cell-linking trait in the ectoderm and endoderm of the Eumetazoa (p. 77), but are completely absent in the Mesozoa. Therefore no support exists for an equation of the epidermis of the Mesozoa with the ectoderm of the Eumeta-

zoa. The uncertainty thus created has led to the use of the more neutral terms ciliated outer cells or sheath cells. In this light, it should be noted that the comparison in the following section between the ciliary rootlets of the epidermal cells and the larval skin cells of parasitic Plathelminthes is being offered with the reservation that the cell layers in question may not be homologous.

Monociliated epidermal cells belong to the ground pattern of the Metazoa, and this plesiomorphous state is consistently realized throughout the Porifera, Placozoa and Cnidaria. In other words, the epidermis of the Mesozoa, with its multiciliated cells, contradicts a basal position within the Metazoa (RIEGER 1976). It would for this reason seem purposeful to search for the sister group of the Mesozoa within the Bilateria among taxa with a multiciliated epidermis. The Bilateria are also recognized by the individualized muscle cells underneath the epidermis, though this is only known to occur in the Orthonectida.

Among the Bilateria, only the larval epidermis of the parasitic Neodermata (Plathelminthes) offers the possibility of a realistic comparison (p. 181). The Plathelminthes have two ciliary rootlets in the ground pattern, a rostral rootlet which is pointed forward, and a caudal rootlet which runs downward or is slanted backward. Only the rostral rootlet is developed in the ciliated epidermal cells of the larvae of the Trematoda, Monogenea, and Cestoda, whereas the caudal rootlet is consistently absent. This absence is indisputably an apomorphy, and lends weight to the proposition that the same number and the same anterior alignment of the ciliary rootlets may be a synapomorphy of the Mesozoa and Neodermata. Argued in this way, support could be found for the opinion that the Mesozoa belong to the parasitic Plathelminthes (STUNKARD 1954). A single congruence that can be interpreted as a synapomorphy with the larvae of the Neodermata is, however, insufficient justification of kinship to a subgroup of the Plathelminthes. There is no argument for the interpretation of the Mesozoa as neotenic Neodermata.

The uncertainty as to how to evaluate the Mesozoa as a whole as well as certain features of the taxon, extends way beyond the Plathelminthes. In fact, no reliable interpretations are available for the absence of zonulae adherentes between epidermal cells, for the absence of a basal lamina underneath the epidermis, or for the lack of sensory and nerve cells.
The adelphotaxon of the Mesozoa within the monophylum Metazoa has not yet been determined.

Porifera – Epitheliozoa

The sponges and a unity comprising all other Metazoa are the highest-ranking subtaxa of the Metazoa (Fig. 18). They fulfill the elementary requirements needed to be interpreted as identically ranking sister groups. Both unities can be justified as monophyla on the basis of autapomorphies. When compared with each other, each exhibits certain plesiomorphous features.

Of the autapomorphies of the Porifera, the sessility and the arrangement of certain cells forming the pinacoderm in the periphery and the choanoderm inside the body are of special interest to us. Septate junctions between cells have been found in the Porifera. The sponges, however, have no zonulae adherentes (HARRISON and DE VOS 1991).
Belt desmosomes (zonulae adherentes) are, on the other hand, an essential autapomorphy of a unity comprising all other metazoans – the Epitheliozoa. In comparison with the sessile sponges, a free-living organism (plesiomorphy) can be postulated for their stem species.

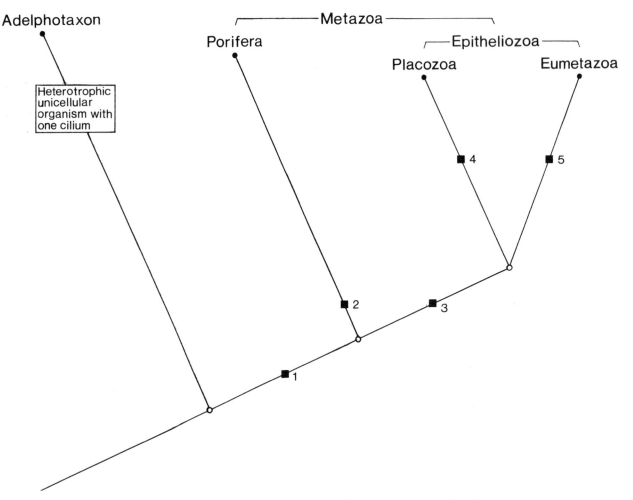

Fig. 18. Diagram of phylogenetic kinship of the Metazoa with the Porifera and Epitheliozoa as the highest ranking adelphotaxa.

Porifera

Sponges do not have sensory cells or nerve cells. Sponges possess no localized gonads. Furthermore, the zonulae adherentes[4] are absent.

In comparison with the other Metazoa, the absence of each of these features is to be considered as a primary absence, and they are accordingly seen as plesiomorphies in the ground pattern of the Porifera.

Sponges do not have a digestive tract, nor do they have glandular cells for the production of exoenzymes to digest food. Sponges are filtering Metazoa that digest food particles intracellularly. A water-conducting or aquiferous system with choanocytes for breathing and nourishment evolved in the stem lineage of the Porifera completely independently of the other Metazoa. This leads us to the derived, unique traits in the ground pattern of sponges.

---

4 Recent findings on comparable cell connections in a Demospongea larva (RIEGER 1994) need to be confirmed within the Porifera.

– Biphasic life cycle with sessile adult and plankton larva (most likely in the form of a coelo-blastula).

– Filter feeding in the adult in combination with sessility.

– Pinacocytes in the form of flat cells. In single-layer arrangements they line the free body surface and basal adhesive surface as exopinacocytes, and the inner ducts as endopinacocytes (Fig. 19).

– Aquiferous system through the body of the sponge with canals and chambers. Inflow through porocytes between the exopinacocytes on the body surface. Outflow of water through localized outlets, the oscula (Fig. 20).

– Choanocytes inside the body with the following substructures (assessments are in parentheses):

Collars made up of numerous microvilli; they surround the proximal part of the cilia in the shape of a collar (apomorphy). Cilia with the 9 x 2 + 2 microtubule pattern (plesiomorphy). Cilia in the proximal part with two wings, known as vanes, which stand out at right angles and laterally reach up to the collar (apomorphy). An accessory centriole (plesiomorphy) next to the basal body of the cilium. Lack of a striated ciliary rootlet (apomorphy; Fig. 21). Choanocytes propel water through the inner aquiferous system. In the ground pattern of the sponges it is possible that several choanocytes were combined together as functional units in the form of collar cell chambers.

– Mesohyl as the interior of the sponge body. Enclosed by pinacocytes and choanocytes.

Ground Pattern

Pinacocytes and choanocytes are the two fundamental characteristics in the cellular organization of sponges. The cellular units they constitute are termed **pinacoderm** and **choanoderm**. However, they do not have a basal lamina or zonulae adherentes between the cells. For this reason, pinacoderm and choanoderm cannot be referred to as epithelia. Due to a lack of specific structural congruences, and also because of their diverging, ontogenetic origins (p. 73), a homologization of the pinacoderm and choanoderm of the Porifera with the ectoderm and endoderm of the Epitheliozoa is not possible.

The **mesohyl** is the body cavity between the pinacoderm and choanoderm (Fig. 20). We have deliberately taken the term mesohyl, which should be used only for the Porifera, because the mesohyl cannot be compared with cavities found in any of the unities of the other Metazoa.

What elements of the few forms of different somatic cells in the mesohyl can be assumed to be part of the ground pattern of the sponges? Only the totipotent stem cells, for which we use the

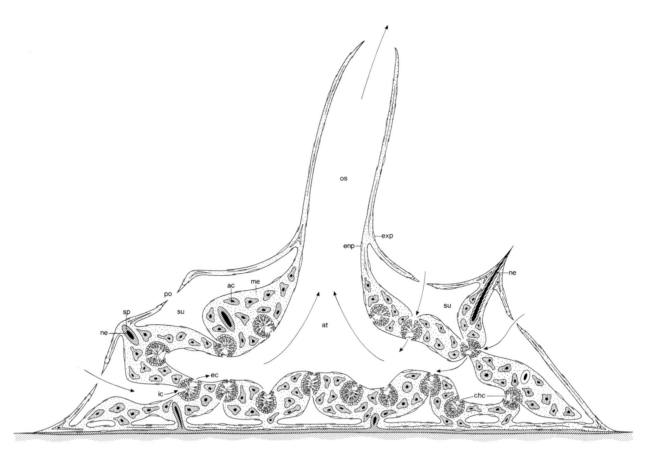

Fig. 19. Porifera. Spongillidae (Demospongea). Organization of a freshwater sponge showing the flow of water through the body. Flat exopinacocytes and endopinacocytes cover the mesohyl. Choanocyte chambers are located between the peripheral subdermal spaces and the central atrium. The choanocytes of the chambers are the motors for the flow of water. Dermal pores form the entrance ways. The water enters into the chambers through incurrent canals. Filtration takes place here. Afterwards the water flows through the excurrent canals into the atrium and leaves the sponge through the osculum. ac = archaeocyte. at = atrium. chc = choanocyte chamber. ec = excurrent canal. enp = endopinacocyte. exp = exopinacocyte. ic = incurrent canal. me = mesohyl. ne = needle. os = osculum. po = dermal pore. sp = spongin. su = sub-dermal space. (Weissenfels 1989)

term archaeocytes, can be adequately justified (Fig. 20). The existence of archaeocytes is, however, a plesiomorphy of the sponges. We hypothesize their adoption from the stem species of the Metazoa. The presence of an extracellular matrix (ECM) with fibrillous collagen in the mesohyl of sponges is also plesiomorphous.

It is possible that contractile cells, which have been found in the vicinity of the dermal pores and the oscular tubes, belong to the ground pattern of sponges. Contractile filaments in the pinaco-cytes of *Ephydatia fluviatilis* consist of actin and myosin (WEISSENFELS 1989), and therefore are comparable with the contractile system of the Eumetazoa. The most immediate assumption of a homology among the actin-myosin systems within the basal Metazoa must, however, be clearly distinguished from the untenable proposition of a homology between individual, contractile cells of sponges and the muscle tissues of the Eumetazoa, which are made up of epithelial or real muscle cells.

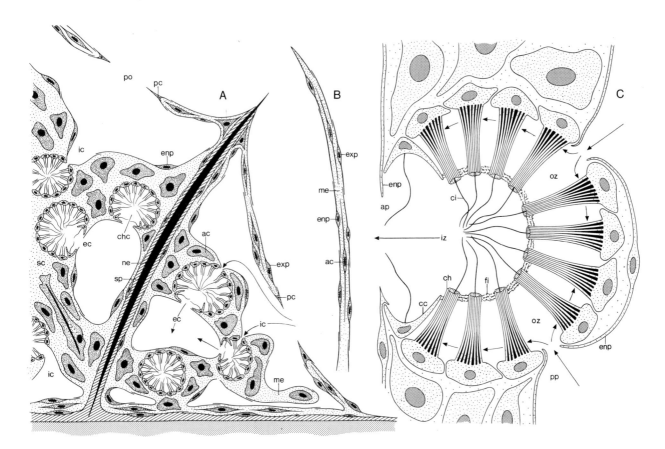

Fig. 20. Porifera. A and B. *Ephydatia fluviatilis.* C. *Spongilla lacustris* (Spongillidae, Demospongea). A. Detail from the periphery of the body with a large siliceous spicule in the soft body and intracellular formation of a new needle in a sklerocyte of the mesohyl. B. Longitudinal section of the wall of the osculum with exopinacocytes and endopinacocytes, and with archaeocytes in the mesohyl. C. Median section of a round choanocyte chamber. Water reaches the outer chamber area (oz) through prosopyles between endopinacocytes (enp). The movement of the cilia pulls the water through slits between the microvilli inside the collar of the choanocytes. The water flows from the apical collar opening further into the inner chamber (iz) and is led by special cone cells through the apopyle into the excurrent canal system. The filter connects the tips of the choanocyte collars forming a network.

ac = archaeocyte. ap = apopyle. cc = cone cell. ch = choanocyte. chc = choanocyte chamber. ec = excurrent canal. enp = endopinacocyte. exp = exopinacocyte. fi = filter. ic = incurrent canal. iz = inner chamber. me = mesohyl. ne = needle. oz = outer chamber zone. pc = porocyte. po = dermal pore. pp = prosopyle. sc = sclerocyte. sp = spongin. (Weissenfels 1989, 1992)

Spongiocytes (spongioblasts), as producers of spongin, and sclerocytes (scleroblasts), which secrete calcium or silica, are autapomorphies of various subtaxa of the Porifera and can be excluded from the ground pattern of the sponges.

Choanocytes, archaeocytes and also pinacocytes take in particulate nourishment from the water in intracellular vacuoles. The **filter mechanism**, which creates the accumulation of particles, is an impressive autapomorphy of the Porifera (Figs. 19,20). In *Spongilla lacustris*, the water passes through several filter structures with decreasing diameter (WEISSENFELS 1992) as it flows through the body of the sponge. The dermal pores (diameter 10-50 μm) hold back unsuitable particles in the periphery. Larger particles are trapped and digested by the endopinacocytes at the beginning of the aquiferous system. Prosopyls, the gateways to the individual collar cell chambers, prevent the passage of particles with a diameter greater than 5-10 μm. The water is pulled out of the outer

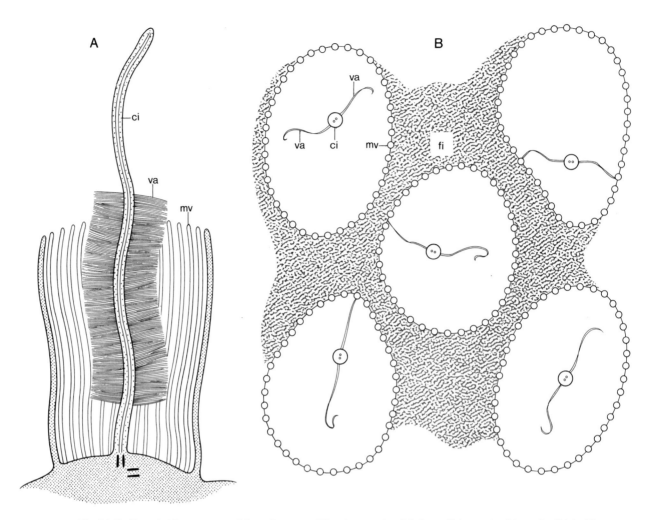

Fig. 21. Porifera. A. Ultrastructure of the collar area of the choanocyte of *Aphrocallistes rastus* (Hexactinellida) with a central cilia, two wing-like formations, and an outer wreath of microvilli. Accessory centriole next to the basal body; the ciliary rootlet is reduced. Cell boundaries of the neighboring choanocytes are absent, since the choanocytes in the chamber of the Hexactinellida are fused to form a syncytium. B. Cross section through a choanocyte chamber of *Spongilla lacustris*. The individual choanocytes are continuously connected by a net-like filter. The filter forms the boundary between the outer and inner chamber. ci = cilium. fi = filter. mv = microvilli. va = vane. (A Mehl and Reiswig 1991; B Weissenfels 1992)

chamber hole, on through fine slits in the basal part of the microvilli collar, into the collar. Finally, the S-shaped movements of the choanocyte cilium propel the water distally out of the collar into the inner chamber and into the system of excurrent canals. It then comes to the final filter station. The individual collars of the choanocytes of a chamber are connected with each other at their distal ends by means of a fine, organic net. This filter net is similar to a glycocalyx. Here, within the collar cell chambers, lies the border between the inward and outward conducting parts of the aquiferous system. The fine particles, which are trapped in the net filter, are phagocytized by special central cells of the chamber using net-shaped, cytoplasmic processes.

The primary absence of localized gonads in the **reproductive system** is particularly noteworthy. Spermatogonia develop from choanocytes, oogonia from choanocytes or archaeocytes. The sperm reach the surrounding waters through the osculum, probably in all sponges. The parallel release of unfertilized eggs and their external fertilization has been observed in only a few Demospongia (FELL 1989). Instead the sperm usually enter ovigerous individuals through the aquiferous system. According to all reports on internal fertilization, individual sperms are taken up by a choanocyte, which then changes into a carrier cell, leaves the choanoderm, and transports the sperm to an egg cell. These events describe a distinctly apomorphous behavior. Considering the observations on the free release of egg cells, internal insemination and the transport of sperm using carrier cells cannot, however, be postulated as part of the ground pattern of the sponges. The current state of knowledge also does not allow us to hypothesize that this mode could be the autapomorphy of a more comprehensive subtaxon within the Porifera.

Three different forms of plankton larva develop through a total, usually equal cleavage. (1) A coeloblastula consisting of a layer of monociliated cells and with a liquid-filled blastocoel has been found among the Calcarea and the Demospongea. (2) In the amphiblastula of some calcareous sponges, only the cells of the front half of the larva are ciliated. Four large, equatorially arranged crosscells are characteristic. The amphiblastula also has a blastocoel. (3) In contrast, the blastocoel of the parenchymula is filled with aciliated blastomeres. This larva form is common amongst the Demospongea, and has also been observed in the Calcarea and in a species of the Hexactinellida (FELL 1989).

Using the out-group comparison described above, the coeloblastula can be assessed as the primary, plesiomorphous larva form of the Porifera due to the fact that the ontogenesis stage of a blastula is very common in the Eumetazoa. We need to be very precise, however, in the next step of argumentation. A homology is possible for the blastula only as a developmental stage, not however, for the coeloblastula as plankton larva of the sponges. There is no justification for postulating a plankton larva as part of the ground pattern of the vagile stem species of the Metazoa. The coeloblastula, as plankton larva, is an autapomorphy of the sponges. It evolved in relation to the transition to sessility in the stem lineage of the Porifera, just as the planula larva of the Cnidaria, in a completely separate process, developed in correlation to the evolution of a sessile polyp.

The comparison with the planula of the Cnidaria clearly shows that, beyond the blastula, no homologization is possible between the ontogenesis of the sponges and the ontogenesis of the other Metazoa (MÖHN 1984 et al.). Both sponge larva and Cnidaria larva attach themselves to the substrate with the animal pole, but their development differs completely after that. Among the Porifera, an invagination of cell material into the body takes place at the place of attachment i.e., the animal pole, from which the choanoderm develops. In the planula larva, invagination results in the formation of the endoderm and a blastoporus directly across from the place of attachment, i.e., on the vegetative pole. The diverging course of ontogenesis further substantiates the argumentation we have formulated. The cell layers of sponges consisting of pinacocytes and choanocytes have nothing in common with the ectoderm and endoderm of the other Metazoa. Moreover the oscular tube of the sponges which develops at the vegetative pole of the larva does not correspond to the blastoporus of the Eumetazoa.

Systematization

In recent phylogenetic literature, the subtaxa Calcarea, Demospongea, and Hexactinellida have been depicted as monophyla. There are two conflicting hypotheses about the kinship relations between the three taxa.

Hypothesis 1

The following hierarchical tabulation is derived from Böger's (1988) systematization of sponges (Fig. 22):

Porifera
Calcarea
Silicea
Demospongea
Hexactinellida

As already mentioned, neither calcareous spicules nor siliceous spicules can be postulated as belonging to the stem species of the Porifera. The two skeletal elements are completely different in construction and formation; they cannot be traced back to a common, basic form. This leads to the following hypothesis: sclerocytes as calcoblasts and sclerocytes as silicoblasts evolved independently of each other in the stem lineages of the sister groups Calcarea and Silicea. In this interpretation, calcareous and siliceous needles become the decisive autapomorphies for the justification of the Calcarea and Silicea as monophyletic unities.

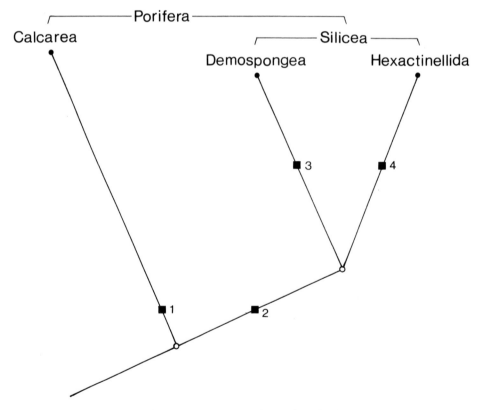

Fig. 22. Diagram of phylogenetic kinship of the highest-ranking subtaxa of the Porifera.

Calcarea – Silicea

Calcarea

■ **Autapomorphy**
(Fig. 22 → 1)

– Existence of sclerocytes which extracellularly secrete calcareous spicules without an axial filament. The individual spicules are made up of several cells. They consist mainly of crystalline $CaCO_3$ with smaller proportions of $MgCO_3$.

Silicea

■ **Autapomorphy**
(Fig. 22 → 2)

– Existence of sclerocytes, which intracellularly secrete a spicule made of SiO_2 around an axial filament made of scleroprotein. A siliceous spiculum is always produced by a single sclerocyte.

Demospongea – Hexactinellida

The Demospongea and the Hexactinellida are sister groups as subordinate taxa of the Silicea. The characteristic spongin is an autapomorphy of the Demospongea, whereas the Hexactinellida are derived by means of their syncytial organization.

Demospongea

■ **Autapomorphy**
(Fig. 22 → 3)

– Existence of spongioblasts which secrete the scleroprotein spongin in the mesohyl.

The collagen of the extracellular matrix is present in all sponges. Collagen is a plesiomorphous feature in the ground pattern of the Porifera. It consists of fibrils around 20 nm in diameter with a simple, striated pattern.

The spongin of the Demospongea is a special form of collagen. It has regularly striated microfibrils that have a diameter of only 4 nm (WEISSENFELS 1989).

The extent of the Demospongea as a monophylum has not yet been clarified. In traditional classifications, taxa without siliceous spicules and taxa without spongin are placed in the Demospongea. As far as justifiable with the methods of phylogenetic systematics, this will force the assumption of partial reduction of the siliceous spicules (Dictyoceratida with bath sponges) and of the spongin (Homoscleromorpha, Tetractinomorpha).

Hexactinellida

■ **Autapomorphies**
(Fig. 22 → 4)

- Syncytial organization through cell fusion. The beam-like tissue forms a system reminiscent of a spider web (HARRISON and DE VOS 1991).

- Absence of pinacocytes.

- Presence of a triaxon of siliceous spicules.

The transformation of the primary, cellular organization into a syncytium encompasses the whole body, including the surface and the choanocyte chambers. Pinacocytes are completely absent. They are replaced by a syncytial dermal membrane. The choanocytes fuse together in the collar cell chambers to form a choanosyncytium.

The triaxon (hexactine) is a six-rayed spicule with three axes meeting at right angles. This structure of the sclerite is usually presented as the basic characteristic of the Hexactinellida. There are, however, simple needles as well. Thus, a triaxon cannot be indisputably postulated for the ground pattern of the Hexactinellida (BÖGER 1988).
Despite this, the Hexactinellida are incontestably justified as a monophylum due to the first two features mentioned above.

Hypothesis 2
In an alternative hypothesis (MEHL and REITNER 1991; MEHL and REISWIG 1991), the Calcarea and Demospongea are postulated as adelphotaxa and are contrasted with the Hexactinellida. This results in the following systematization.

Porifera
Hexactinellida
N.N.
Calcarea
Demospongea

This hypothesis also assumes that neither calcareous nor siliceous sclerites were present in the ground pattern of the sponges. At the same time, however, it assumes that siliceous spicules with axial filaments evolved separately in the stem lineage of the Hexactinellida and in the stem lineage of the Demospongea. Emphasis is placed on a small difference in the cross section of the axial filament. This cross section is reputed to be square-shaped in the Hexactinellida and triangular in the Demospongea. The divergence cannot be seen as important, particularly since the alternative is not constant. The cross section of the axial filament in the Spongillidae, which is a subtaxon of the Demospongea, has been shown to be hexagonal (WEISSENFELS 1989).
The principle of parsimony dictates that we give precedence to the assumption that siliceous sclerites with an axial filament evolved only once. Since this assumption does not conflict with the assessment of other features in hypothesis 1, I favor the phylogenetic systematization of the Porifera presented in the first of these two hypotheses.

Epitheliozoa

■ **Autapomorphies**
(Fig. 18 → 3)

– Epithelial linking of cells by means of the development of zonulae adherentes.

– Differentiation of two epithelial layers: (1) dorsal epithelium of the Placozoa = ectoderm of the Eumetazoa; (2) ventral epithelium of the Placozoa = endoderm of the Eumetozoa.

– Glandular cells in the ventral epithelium that produce exoenzymes for extracellular digestion.

A new type of connecting structure between neighboring cells – the zonulae adherentes (Fig. 23C) – evolved in the stem lineage of the Epitheliozoa. They surround the individual cells of a specific cell layer as a belt or band. The plasma membranes of neighboring cells are arranged parallel to each other in the area of the belt, and the intercellular space contains filamentary material. Bundles of actin filaments run just beneath the cell membrane in the area of the belt. The zonulae adherentes create a cluster of cells which react together and which are capable of common changes in form.

The word **epithelium** can be used as a clearly defined term if it refers only to cell layers with zonulae adherentes (Bartolomaeus 1993a). This is why we have decided not to use the term for the pinacoderm and choanoderm of sponges.

Zonulae adherentes are found in the adelphotaxa Placozoa and Eumetazoa. Together, they form a monophyletic unity that we name the Epitheliozoa.

The development of zonulae adherentes was the starting point for the evolution of two structurally and functionally different epithelia. The first step of this process can be seen in the Placozoa. A dorsal layer of flat cells faces the water and offers protection from the surrounding environment. The ventral epithelium with ciliated cylindrical cells provides the propulsion for moving on the substrate. Food intake also takes place on the ventral side. Aciliated glandular cells, embedded between the ciliated cells, produce exoenzymes for extracellular digestion.

The existence of these glandular cells in the Placozoa permits the hypothesis that the ventral epithelium is homologous to the endoderm of the Eumetazoa. This leads to the assumption of equivalence between the dorsal epithelium of the Placozoa and the ectoderm of the Eumetazoa.

Placozoa – Eumetazoa

When compared with each other, the Placozoa show original traits with minimal somatic differentiation in only four different cell forms. These are the flat cells of the dorsal epithelium, the cylindrical and glandular cells in the ventral epithelium, and special fiber cells in the middle layer. The latter have no equivalent in the Eumetazoa, and are, for this reason, hypothesized to be a unique trait of the Placozoa. Completely new cell forms evolved in the stem lineage of the Eumetazoa – sensory cells, nerve cells, and epithelial muscle cells.

Placozoa

■ **Autapomorphies**
(Fig. 18 → 4)

- Contractile fiber cells in the intermediate layer of the body.

- Cilia of the ventral cylindrical cells have two horizontal ciliary rootlets in addition to the vertical main rootlet (BARTOLOMAEUS 1993a).

- Absence of extracellular matrix and collagen (BARTOLOMAEUS 1993a)

Trichoplax adhaerens F.E. Schulze, 1883 from the littoral zone of warm oceans is the only sufficiently described species of the taxon Placozoa (GRELL 1971). The 2–3-mm-large, dorsoventrally compressed organism has no axes of symmetry. The animal crawls on the substrate, constantly changing its contours in an "amoeba-like" manner (Fig. 23A). The cells of the dorsal and ventral epithelium represent the original monociliated state of the Metazoa. Other plesiomorphies consist of an accessory centriole next to the basal body and a long, striated ciliary rootlet perpendicular to the cell surface. Basal lamina are absent.
There are clear differences between the epithelia (Fig. 23E).

The **dorsal epithelium** is composed of very flat cells with a height of only a few μm. Light-refracting, lipid spheres of unknown origin are present in the epithelium. Cylindrical ciliated cells with a locomotory function are the main components of the **ventral epithelium**. Next to the vertical rootlet they each have two additional short and striated ciliary rootlets, which arise horizontally from opposite sides of the basal body (RUTHMANN, et al. 1986; Fig. 23D). Aciliated glandular cells are located between the cylindrical cells.
The dorsal and ventral epithelium border the intermediate layer, a liquid-filled space with a network of **fiber cells** which have no equivalent in the other Metazoa (apomorphy). Long processes connect the fiber cells with each other and with the epithelia; actin filaments in their cytoskeleton cause rapid changes in body shape. No evidence of a collagenous extracellular matrix has been found.
Food intake occurs mainly through the ventral epithelium. Microorganisms are predigested by the exoenzymes of the glandular cells; the breakdown products enter the cylindrical cells by means of pinocytosis. In addition, phagocytosis of algae cells and yeast cells by the fiber cells has been observed (GRELL and RUTHMANN 1991).
Trichoplax adhaerens can be kept for years in seawater aquaria, where they actively undergo vegetative propagation (binary fission, formation of multicellular swarmers by budding; Fig. 23 B). Unfortunately, however, only insufficient knowledge of the sexual processes is available. Egg cells and aciliated sperm have been observed in the intermediate layer (GRELL 1972; GRELL and BENWITZ 1974, 1981), but not the unification of gametes. The separation of a fertilization membrane and the first steps of a complete and equal cleavage have been observed.
Years of clone breeding and the indications available pointing to sexual reproduction speak against the suspicion that *Trichoplax adhaerens* could be a larval form of unknown affiliation. We have to be careful with phylogenetic interpretations, however, until a more complete picture of the life cycle of *Trichoplax adhaerens* has been established. This holds true for the interpretation of

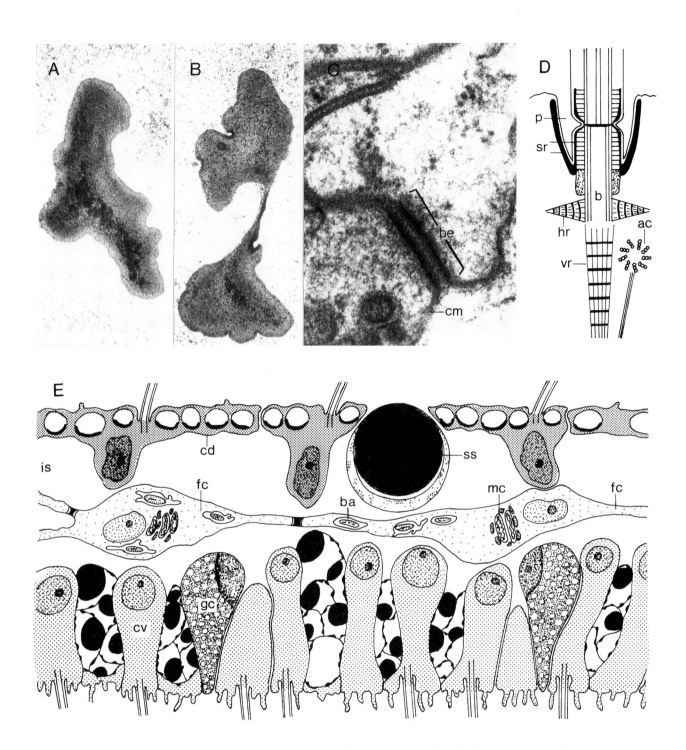

Fig. 23. *Trichoplax adhaerens* (Placozoa). A. Crawling organism with amoeba-like outline. B. Vegetative reproduction in the form of binary fission. C. Cross section of a zonula adherens and the cell membrane. D. Cilium of the ventral epithelium. Basal apparatus in a ciliary pit. Basal body with two horizontal and one vertical ciliary rootlet. E. Longitudinal section showing histological construction based on electron microscopic examinations. ac = accessory centriole. b = basal body. ba = bacterium. be = belt desmosome (zonula adherens). cd = monociliated cell of the dorsal epithelium. cm = cell membrane. cv = monociliated cell of the ventral epithelium. fc = fiber cell. gc = aciliated glandular cell of the ventral epithelium. hr = horizontal ciliary rootlet. is = intermediate layer. mc = mitochondrium. p = ciliary pit. sr = supporting rod. ss = shiny sphere. vr = vertical ciliary rootlet. (A, B, and C originals from A. Ruthmann, Bochum; D Ruthmann et al. 1986; E Grell 1981)

79

fiber cells as an autapomorphy of the Placozoa, and also for the hypothesis that the absence of an extracellular matrix and collagen represents a derived state.

Eumetazoa

■ **Autapomorphies**
(Fig. 18 → 5)

– Ectoderm: epithelium which completely surrounds the body.

– Endoderm: epithelium which surrounds a gut cavity inside the body. In the ground pattern of the Eumetazoa, the gut cavity has only one opening (plesiomorphy) that functions simultaneously as mouth and anus.

– Monociliated sensory cells in the epidermis. Each sensory cell has a stiff cilium (sterocilium) and sends out slender neurites from its base.

– Nerve cells at the base of both epithelia. The nerve cells or ganglion cells join together with their processes forming a diffuse network of nerves.

– Epithelial muscle cells: presence of muscle fibers in the base of ectodermal and endodermal cells.

– Gap junctions between neighboring cells (BARTOLOMAEUS 1993a).

Ectoderm, endoderm, and mesoderm have been the central terms in the morphology of the Eumetazoa for some time now, and we would, for this reason, like to begin this section by clearly defining these terms. We will then ask the question as to how far the defined meaning proves useful in phylogenetic systematics.

Ectoderm and endoderm were coined as topographical terms to label the two body layers of adult hydrozoans (ALLMAN 1853). When applied to all Eumetazoa, the following definitions can be formulated for adults (BARTOLOMAEUS 1993a):

Ectoderm
"The outer epithelium which completely encloses the body."

Endoderm
"The inner epithelium which lines a hollow cavity permanently surrounded by the body. This cavity communicates with the outer environment by means of at least one opening."

The ectoderm is often referred to as the epidermis, and the endoderm as the gastrodermis.

Phylogenetic systematics hypothesizes that the ectoderm and endoderm of all eumetazoans form homologous body layers. Ectoderm and endoderm evolved in the stem lineage of the Eumetazoa from a disk shaped dorsal and ventral epithelium, comparable to the characteristics found in the Placozoa today. The decisive evolutionary step was the sac-shaped indentation of the ventral epithelium, forming an inner, hollow cavity for the optimization of extracellular enzymatic digestion. A process of this kind unavoidably leads to expansion of the dorsal epithelium to an ectoderm which surrounds the body.

In comparison with the epithelia of the Placozoa, the ectoderm and endoderm of the Eumetazoa are interpreted as apomorphous features. Based on the most parsimonious assumption of a unique evolution of ectoderm and endoderm, they form synapomorphies of the sister groups Cnidaria and Acrosomata, which makes them autapomorphies of the Eumetazoa.

Mesoderm

The word mesoderm was also used as a topographical term in the last century (LANG 1891; KORSCHELT and HEIDER 1892, 1895). We define mesoderm as follows:

"All tissue between the ectoderm and endoderm embedded in the extracellular matrix (ECM) – regardless of which of these two body layers it originates from" (RUPPERT 1991; BARTOLOMAEUS 1993a).

According to this definition, the mesoglea with cells of certain Cnidaria and the Ctenophora, as well as the divergent tissues found in the intermediate layer of the Bilateria, are mesoderm. There is, however, a fundamental difference in comparison with the ectoderm and the endoderm. A consistent homologization of the topographically defined mesoderm is not possible within the Eumetazoa. The mesoglea of the Cnidaria and the Ctenophora originated independently of each

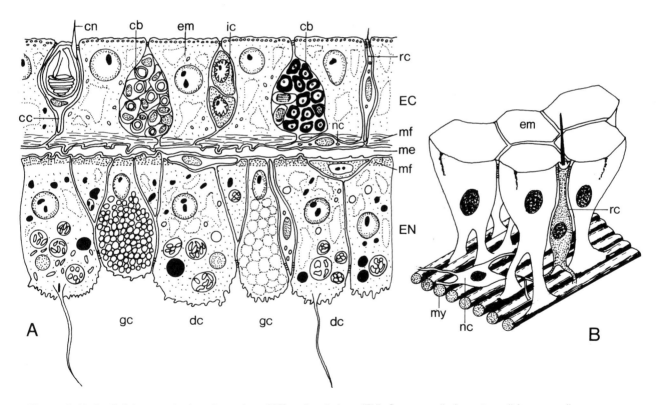

Fig. 24. Cnidaria. Cellular organization of ectoderm (EC) and endoderm (EN). Sensory cells (receptor cells), nerve cells, and epithelial muscle cells form autapomorphies in the ground pattern of the Eumetazoa. A. Section of the body wall of *Hydra*. B. Epithelial muscle cells, a receptor cell, and nerve cell of the ectoderm of a polyp of the Hydrozoa. cb = cnidoblast. cc = cnidocyte. cn = cnidocil. dc = digestive cell. em = epithelial muscle cell. gc = gland cell. ic = interstitial cell. me = mesogloea. mf = muscle fibrils. my = muscle fiber. nc = nerve cell. rc = receptor cell. (A Kozloff 1990; B Werner 1993)

other (p. 87, 107); the mesoglea of the Ctenophora has nothing in common with the manifestation of the mesoderm in the ground pattern of the Bilateria (p. 112).

Sensory Cells – Nerve Cells – Muscle Cells

The evolution of sensory cells, nerve cells, and muscle cells in the stem lineage of the Eumetazoa were essential steps in the phylogenesis of the Metazoa. We can postulate those manifestations of sensory, nerve, and muscle cells to be elements of the ground pattern of the Eumetazoa which are original traits within the Cnidaria. The evolution of sensory systems began with simple, monociliated sensory cells in the epidermis. The evolution of conduction of stimuli by means of nerve cells began with a loose network of nerves in the base of the ectoderm and endoderm. A coordination center (brain) did not yet exist in the stem species of the Eumetazoa; it first evolved in the stem lineage of the Bilateria. An efficient locomotory apparatus developed in the simplest way possible; muscle fibers of actin and myosin filaments evolved in the base of normal ectoderm and endoderm cells which, with an expansion of their functions, became epithelial muscle cells (Fig. 24).

Cnidaria – Acrosomata

What is the relationship between the three taxa of the Eumetazoa – Cnidaria, Ctenophora and Bilateria?

The attempt to link the Cnidaria and Ctenophora with the help of "the medusae" and to create a monophylum for these two unities bearing the traditional name Coelenterata (Ax 1989c) has proved to be unsuccessful. The ultrastructure has revealed two apomorphies which are independent of each other and which only the Ctenophora and the Bilateria have in common.

The first feature is the complex of acrosome and subacrosomal substance (perforatorium) in the head of the sperm. Let us first mention sperm which have several small acrosomal vesicles. Sperm of this kind appear in the Porifera as well as in the Cnidaria and are thus hypothesized as being part of the ground pattern of the Metazoa (p. 54). On the other hand, the sperm of the Ctenophora and Bilateria have a uniform acrosome, and in addition, a subacrosomal substance called perforatorium (Fig. 14). This apomorphous congruence can be proved without conflict to be a synapomorphy of the Ctenophora and Bilateria. Sperm with an acrosome and subacrosomal substance evolved once in a stem lineage common only to the Ctenophora and Bilateria (EHLERS 1993).

The second congruence results from the structure of the muscle cells. The Cnidaria have the plesiomorphous feature of monociliated epithelial muscles. In contrast to this, the musculature system of the Ctenophora, like that of the Bilateria, consists exclusively of muscle cells which are not epithelially arranged. Myocytes in the Ctenophora are present in the form of individual, independent cells in the ectoderm and endoderm or as a muscle syncytium in the mesoglea. The existence of individualized myocytes can be assessed as another synapomorphy of the Ctenophora and Bilateria, i.e. as an apomorphy that evolved once from the primary state of the epithelial muscle cells (BARTOLOMAEUS 1993a).

In this interpretation, sperm with uniform acrosome and perforatorium as well as real muscle cells become the autapomorphies of a taxon Acrosomata that comprises the sister groups Ctenophora and Bilateria (Fig. 25).

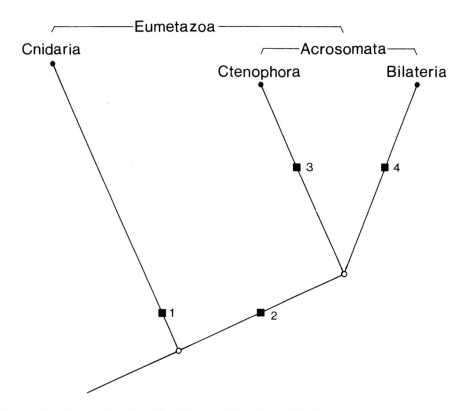

Fig. 25. Diagram of phylogenetic kinship of the highest-ranking subtaxa of the Eumetazoa.

A hypothesis which has been formulated based on RNA sequence analyses, according to which the Ctenophora are supposed to be the sister group of all other Epitheliozoa (Placozoa, Cnidaria, Bilateria) (Wᴀɪɴʀɪɢʜᴛ et al. 1993), is not unifiable with the systematization introduced above. The named congruences would then not only have to be proven as convergences. A further consequence would be the assumption of a twice independent evolution of sensory cells, nerve cells, and muscle cells in the Ctenophora, and in a unity of the Cnidaria and Bilateria. There is no justification for this hypothesis whatsoever.

Cnidaria

■ **Autapomorphies**
(Fig. 25 → 1)

– Sessile polyp with radial symmetry. No septa.

– Biphasic life cycle with a planula larva in the pelagial.

– Cnidae (nematocysts) as products of individual cells, the cnidocytes or cnidoblasts.

Ground pattern

Justification of the Cnidaria as a monophylum is possible due to the existence of cnidae, the element which gives the unity its name. There is no dissent on this point. Opinion does, however, differ as to whether the stem species of the Cnidaria was a polyp, a medusa, or may have been even a species with both a polyp and a medusa.

The stem species of the Cnidaria as a polyp
Since the possibility of a homology between the medusal stage of the Cnidaria and the medusoidal organization of the Ctenophora cannot be further pursued (p. 82), let us study the following facts without the burden of a preconceived hypothesis. All of the Anthozoa, Hydrozoa and Rhopaliophora (Cubozoa + Scyphozoa) have a polyp, but only the Hydrozoa and Rhopaliophora also have a medusa.

This clearly indicates the existence of a polyp in the ground pattern of the Cnidaria. At the same time, we will postulate that the planula larva is a part of the ground pattern. The larva originated as a propagative stage during the evolution of a sessile polyp in the stem lineage of the Cnidaria. And what about the "medusa"? To answer this question, one has to decide whether the absence of a medusa in the Anthozoa is a primary state or a secondary loss. The second alternative is a priori the more risky hypothesis. The possibility of the secondary absence of a medusa would be conclusive only if the Anthozoa could be systematized as a subordinate subtaxon of the Hydrozoa, the Cubozoa, or the Scyphozoa, because their very existence requires that medusae be placed in the ground pattern of the Hydrozoa, Cubozoa, and Scyphozoa. The Anthozoa, however, cannot be systematized in this manner, because several of their features are clearly more primitive than in all other Cnidaria. Without going into further detail at this point, reference is made to the mechanoreceptor of the Cnidocytes for the discharge of nematocysts. Compared with the complex apomorphous cnidocil of the Hydrozoa, Cubozoa, and Scyphozoa, the Anthozoa have a simple cilium with a striated ciliary rootlet as a plesiomorphous feature (Fig. 27).

Today, the Anthozoa are justifiably interpreted as the adelphotaxon of the other Cnidaria (Fig. 26), combined under the name Tesserazoa (v. SALVINI-PLAWEN 1978). There are no facts and no convincing hypotheses in favor of the secondary absence of a medusa. Strict rational argumentation proves the Anthozoa to be Cnidaria primarily without medusae.
The conclusion to be drawn is that only a polyp, in combination with the planula larva, can be assumed to represent the ground pattern of the Cnidaria.

The Structure of the Polyp of the Stem Species

We will now discuss in detail the special features of the polyp in the ground pattern of the Cnidaria.

T h e   S o l i t a r y   P o l y p . The Anthozoa and the Hydrozoa have colonies of polyps linked to each other as well as solitary polyps; the Cubozoa and Scyphozoa exist only as individual polyps. In other words, all the taxa mentioned have solitary polyps. Thus, the origin of polyp colonies by means of budding must be identified as an apomorphy within the Anthozoa and the Hydrozoa. A solitary polyp features in the ground pattern of the Cnidaria.

T h e   P r i m a r y   A b s e n c e   o f   S e p t a . The polyps of the Cubozoa and Hydrozoa have an undivided coelenteron. Septa of endodermal origin project out into the gut lumen in the Anthozoa and the Scyphozoa. There is no sound argument to prove that the aseptal state of the Cubozoa and Hydrozoa is a secondary phenomenon. Based on the principle of parsimony, we postulate a radially symmetrical polyp without septa for the stem species of the Cnidaria.

This hypothesis is also supported by the fundamental structural differences in the septa of the Scyphozoa and Anthozoa. All Scyphozoa have four septa with an ectodermal muscular system which enters into the septa along ectodermal funnels at the base. The Anthozoa, however, always have a greater number of septa with lateral muscle bands of endodermal origin.

These diverging features cannot be traced back to one common basic form; they are interpreted to be non-homologous formations (SCHUCHERT 1993). In other words, we must hypothesize an independent evolution of septa in the stem lineages of the Anthozoa and the Scyphozoa.

H o l l o w   T e n t a c l e s . Tentacles with a hollow cavity in the endoderm characterize the Anthozoa; they also appear in some Hydrozoa. Solid tentacles with an alumenal axis made of endodermal cells appear in the majority of the Hydrozoa as well as in all Cubozoa and Scyphozoa. It is difficult to imagine that hollow tentacles developed from a state with a solid endodermal axis. We conclude that hollow tentacles are a part of the ground pattern of the Cnidaria (SCHUCHERT 1993).

E n d o d e r m a l   G o n a d s . The gonads of the anthozoans develop in the endoderm of the polyp. On the shift of sexual reproduction to the medusa, the gonads remain in the endoderm in the Cubozoa and the Scyphozoa; only in the Hydrozoa are they found mainly in the ectoderm. Endodermal gonads must be postulated for the stem species of the Cnidaria.

Evolution of Medusae

In the absence of a medusa in the ground pattern of the Cnidaria (p. 84), let us look for a logical explanation for the evolution of the medusae within the Cnidaria.

The three taxa that have medusae, Hydrozoa, Cubozoa and Scyphozoa, form the sister group of the primary amedusal Anthozoa united under the name Tesserazoa. Within the Tesserazoa, the Cubozoa and Scyphozoa are the adelphotaxa of a subordinate monophylum Rhopaliophora. On the basis of this systematization, which will be justified in detail later on, there are at least three possibilities for the evolution of medusae; these have all been discussed in recent phylogenetic literature.

(1) A medusa originated once in the stem lineage of the Tesserazoa (PETERSEN 1979, SIEWING 1985).
(2) Medusae evolved twice within the Cnidaria – once in the stem lineage of the Hydrozoa and once in the stem lineage of the Rhopaliophora (THIEL 1966; v.SALVINI-PLAWEN 1978).

(3) Medusae originated three times independently of each other, in the Hydrozoa, the Cubozoa and the Scyphozoa (WERNER 1984).

The last of these hypotheses is untenable. Two conspicuous apomorphous congruences exist between the cubomedusae and the scyphomedusae: the complex rhopalia on the border of the bell and the endodermal gastric filaments in the stomach. Their assessment as synapomorphies leads to the postulation of a medusa as a part of the ground pattern of the Rhopaliophora.

The principle of parsimony dictates that we attempt to justify a single evolution of medusae for the Hydrozoa, Cubozoa and Scyphozoa. This, however, is difficult. Let us look at the different ways medusae are formed in the Scyphozoa and Hydrozoa. The Scyphozoa are characterized by the terminal budding of medusae in the form of ephyrae. In the majority of the Hydrozoa, however, the medusae are formed through lateral budding from the polyp. Even if both types of medusae

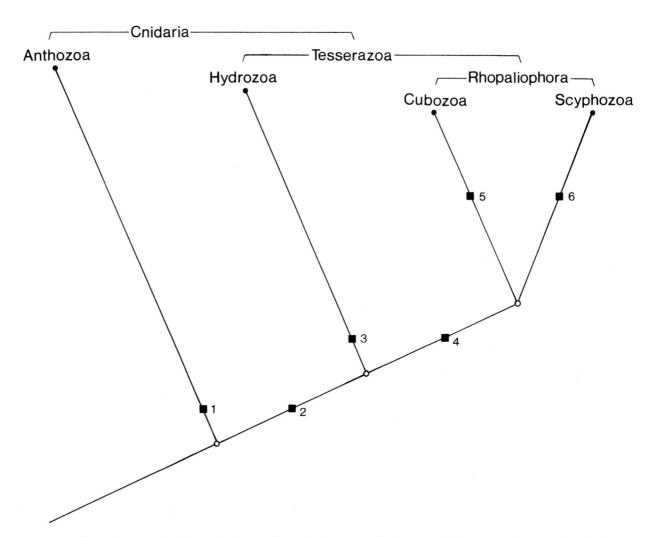

Fig. 26. Diagram of phylogenetic kinship of the Cnidaria. The traditional division into three "equal-ranking classes", Hydrozoa, Scyphozoa, and Anthozoa, must yield to a systematization in which the Anthozoa and the Tesserazoa (all other cnidarians) are positioned as adelphotaxa. Within the Tesserazoa, the Hydrozoa form the sister group of the Rhopaliophora; the Cubozoa and the Scyphozoa are united under this name.

formation possibly could be derived from a primary complete metamorphosis of polyps into medusae (p. 94, 102), they nevertheless signify two completely separate paths of asexual origination. This supports the second hypothesis, which is that the medusae evolved separately in the Hydrozoa and Rhopaliophora.

Moreover, the phylogenetic systematization of the Cnidaria presented in the following is independent of the question as to whether the medusae evolved once, or convergently twice, within the Cnidaria. The systematization is based on the assessment of features for which this problem is irrelevant.

Mesoglea

The mesoglea is the body layer between the ectoderm and endoderm of the Cnidaria. It is made up of a more or less extensive ECM with fibrillar collagen. Among the hydropolyps and some of the scyphopolyps, the mesoglea forms only a thin supporting lamella between the epithelia; it is strongly developed in the anthopolyp, however, and in the medusae, the mesoglea becomes a voluminous layer between ectoderm and endoderm.

Cells of ectodermal origin are present in the mesoglea of the Anthozoa and the majority of the Scyphozoa. On the other hand, the mesoglea is acellular in the Scyphozoa unity of the Coronata (polyp), in the Cubozoa (polyp and medusa), and in the Hydrozoa (polyp and medusa). This distribution does not seem to enable us to reliably justify a cellular mesoglea in the ground pattern of the Cnidaria, and to explain the absence of cells in the Hydrozoa as a derived feature (SCHUCHERT 1993). It is more likely that an independent migration of ectodermal cells into the mesoglea occurred in the Anthozoa and within the Scyphozoa.

Systematization

Cnidaria
Anthozoa
Tesserazoa
Hydrozoa
Rhopaliophora
Cubozoa
Scyphozoa

Anthozoa – Tesserazoa

Let us exclude the medusa from our justification of an adelphotaxa relation between the Anthozoa and the Tesserazoa, since the problem of a single or double evolution of medusae within the Cnidaria would not yet seem to have been completely clarified.

The primary amedusal Anthozoa show autapomorphies in that they have septa with endodermal muscle bands and an ectodermal pharynx with siphonoglyphs.

The Anthozoa, however, are clearly more primitive in two of their features than all other Cnidaria. The first is a molecular feature, which we will present with its alternatives. The conformation of the mitochondrial DNA molecule was analyzed in 48 species of Cnidaria (BRIDGE et al. 1992). The 17 representatives of the Anthozoa all have a ring-shaped mtDNA. All 31 species from the taxa Hydrozoa, Cubozoa and Scyphozoa are, on the other hand, characterized by a linear mtDNA structure. The polarity of this alternative feature is irrefutably established by means of the outgroup comparison.

The Ctenophora, as with all other Eumetazoa that have been examined for this feature, always have a ring-shaped mtDNA. This proves that the Anthozoa have the plesiomorphy of the alternative, and that the linear form of the molecule in the Tesserazoa is the apomorphy. The large number of species examined provides enough evidence to justify the assumption that a linear mtDNA molecule is an autapomorphous feature in the ground pattern of the monophylum Tesserazoa.

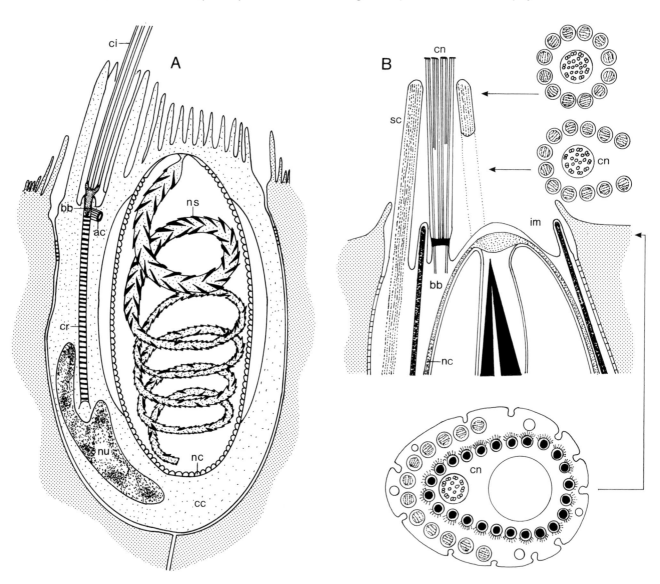

Fig. 27. Apparatus for the release of cnidae, comparison of the Anthozoa (A) with the Tesserazoa (B). A. Diagram of the cnidocyte in the Octocorallia (Anthozoa). The cell has a normal cilium with basal body, accessory centriole and a long rootlet. B. *Craspedacusta sowerbii* (Hydrozoa). Distal portion of the cnidocyte with cnidocil apparatus as a result of transformation of the cilium. The stiff cnidocil has a basal body, but does not have a rootlet or an accessory centriole. ac = accessory centriole. bb = basal body. cc = cnidocyte. ci = cilia. cn = cnidocil. cr = ciliary rootlet. im = inner microvilli. nc = cnidae wall. ns = cnidae tube. nu = nucleus. sc = stereocilia. (A Schmidt and Moraw 1982; B Holstein and Hausmann 1988)

The second feature results from an alternative in the ultrastructure of the mechanoreceptor of the cnidocytes involved in the discharge of nematocysts (Fig. 27).

The Hydrozoa, Cubozoa, and Scyphozoa all have a stiff cnidocil, which is interpreted to be the result of the transformation of a regular cilium (SCHUCHERT 1993). The cnidocil in the Hydrozoa has been analyzed in detail (HOLSTEIN and HAUSMANN 1988). The regular 9 x 2 + 2 microtubule pattern of mobile cilia is absent. Nine double tubules surround a variable group of individual microtubules in the basal region of the cnidocil, whose numbers increase the closer they are to the distal region. The cnidocil has a regular basal body, but no ciliary rootlet, and there is also no accessory centriole. The cnidocil apparatus includes an inner ring of short microvilli, and an outer ring of long microvilli (stereocilia) which surround the cnidocil forming differentiations in the cnidocyte wall. In the distal region of the cnidocyte, a fibrillar collar surrounds the base of the cnidocil.

In contrast, the Anthozoa have no cnidocil. An original, mobile cilium (9 x 2 + 2 microtubule pattern) with a long, striated ciliary rootlet appears in its place, and an accessory centriole is present here as well. Microvilli of the cnidocyte and microvilli of the neighboring epidermal cells surround the cilia.

In the alternative described, the sensory apparatus of the cnidocytes of the Anthozoa is without any doubt the plesiomorphous state, and its formation into a stiff cnidocil is the apomorphy. This enables us to present a second characteristic which can be interpreted as an autapomorphy for the justification of the Tesserazoa as a monophylum.

Anthozoa

■ **Autapomorphies**
(Fig. 26 → 1)

– Endodermal septa with lateral endodermal muscle bands
(consisting of epithelial muscle cells) and terminal mesenterial filaments.

– Ectodermal actinopharynx, which runs from a flat, oral disk and sinks deep into the coelenteron. Mouth and pharynx are flattened at the sides, and the lumen of the pharyngeal canal is, as a result, cleft-shaped.

– Siphonoglyphs with cilia in the recesses of the pharynx; they transport water into the coelenteron.

– Bilateral symmetry. The sagittal plane which runs through the diametrically largest part of the pharynx divides the polyp into two symmetrically equal halves.

The **bilateral symmetry of the Anthozoa** is the result of the evolution of a laterally compressed pharynx with siphonoglyphs, as well as of the evolution of septa with muscle bands on one side only (Fig. 28). Certain Hexacorallia (Actiniaria, Madreporaria) also exhibit **biradial symmetry** or disymmetry. As a result of a biradial arrangement of the muscle bands, the polyp can be dissected into two symmetrically identical halves along a transversal plane perpendicular to the sagittal plane (Fig. 28).

Bilateral symmetry and biradial symmetry are characteristics of the Anthozoa or parts of the Anthozoa which evolved within the monophylum Cnidaria from a primary radial symmetry. The bilateral symmetry of the Bilateria and the biradial symmetry of the Ctenophora thus only share the terms used to describe them. They evolved independently of the symmetry found among the Anthozoa and independently of each other in the stem lineages of the Bilateria and of the Ctenophora.

Hexacorallia and Octocorallia, the two highest ranking unities in traditional classifications of the Anthozoa, are most likely adelphotaxa.

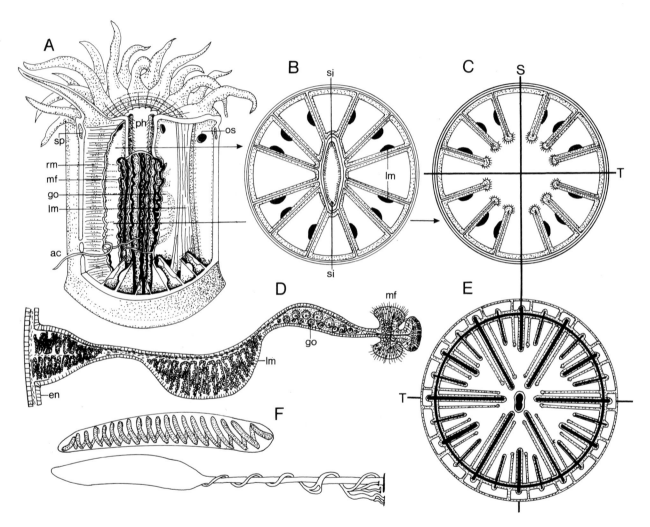

Fig. 28. Hexacorallia (Anthozoa). A. Diagram showing the organization of a polyp of the Actiniaria (sea anemones). Sector cut out. Right: complete septum with endodermal muscle band (longitudinal muscle). Left: incomplete septum with radial muscles which the pharynx does not reach. B. Actiniaria. Cross section at the level of the pharynx. C. Actiniaria. Cross section at the level of the central stomach. (S = sagittal plane, T = transversal plane). D. Actiniaria. Cross section through a septum with muscle band and terminal mesenterial filament. E. Diagram of the skeletal formation in the Madreporaria. With the formation of calcareous septa (sclerosepta) between the soft septa (sarcosepta), the stone corals follow the symmetrical behavior of the actinias. F. Spirocyst of *Peachia hastata* (Actiniaria). Dormant state and after discharge. ac = acontia. en = endoderm. go = gonad. lm = longitudinal muscle. mf = mesenterial filament. os = ostium. ph = pharynx. r= radial muscle. si = siphonoglyph. sp = sphincter. (A-D, F Werner 1993; E Chevalier 1987)

The **Hexacorallia** can be justified as a monophylum on the basis of the spirocysts which are characteristic only to them. These are cylindrical, thin-walled capsules each with a spirally wound, main thread containing a swell thread. After discharge, the swollen thread surrounds the tube in a counter-clockwise spiral and dissolves into a mucous secretory network (Fig. 28).

Attempts to phylogenetically systematize the Hexacorallia in reference to the cnidocytes (SCHMIDT 1972, 1974) have not yet yielded a reliable picture.

The eight septa of the **Octocorallia** could be an autapomorphy of this unity if the formation of six septa as is seen in the Antipatharia (Hexacorallia), depicts a ground pattern trait of the Anthozoa (HENNIG 1994). In our opinion, the pinnate character of the eight tentacles is a uniquely derived trait (Fig. 29) which, among the Cnidaria, occurs only in the Octocorallia.

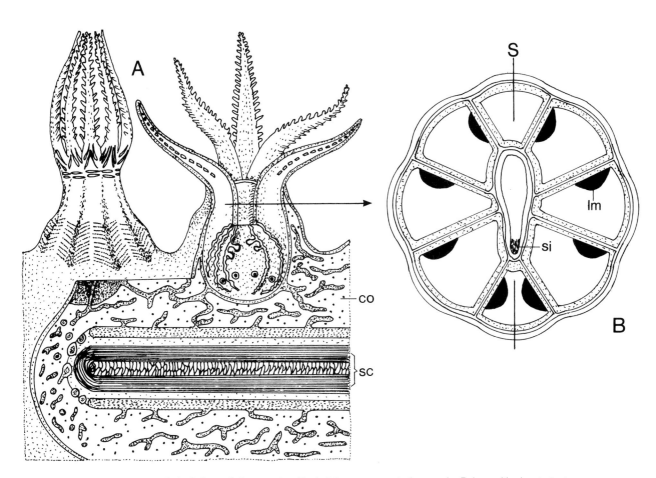

Fig. 29. Octocorallia (Anthozoa). A. Colony of *Gorgonaria* with skeleton composed of gorgonin. Polyps with pinnate tentacles. B. Cross section of an Octocorallia polyp at the level of the pharynx. (S= sagittal plane). co = coenenchym with endodermal canals. lm = longitudinal muscle. sc = skeleton. si = siphonoglyph. (Werner 1993)

Tesserazoa

■ **Autapomorphies**
(Fig. 26 → 2)

– Cnidocil apparatus with a stiff cnidocil as the evolutionary transformation of a cilium.

– Linear arrangement of the mitochondrial DNA molecule.

– Microbasic eurytele.

After examining the autapomorphies of the Tesserazoa (p. 88), we can now move on to a justification of the sister group relationship Hydrozoa – Rhopaliophora (Cubozoa + Scyphozoa).

Hydrozoa – Rhopaliophora

Hydrozoa and Cubozoa have simple polyps without endodermal septa. Neither facts nor conclusive arguments from preconceived hypotheses are available to prove that this condition is the result of an evolutive loss of septa. A polyp with a uniform coelenteron is a plesiomorphy in the ground pattern of the Tesserazoa.

Either a complete medusa (p. 87), or at least group specific characteristics of the medusa (the velum of the Hydrozoa and the marginal rhopalia in the Rhopaliophora) amount to autapomorphies of the adelphotaxa.

Hydrozoa

■ **Autapomorphy**
(Fig. 26 → 3)

– Medusa with an ectodermal velum.

The hydromedusa with a velum or the velum alone can be assumed to be a derived feature of the Hydrozoa. The velum is an ectodermal formation (Fig. 30). It grows from the rim of the medusa as a diploblastic leaf out of the exumbrella and subumbrella far beneath the bell, and forms an aperture that increases the force with which water is expelled.

Fig. 30. Hydrozoa. A. Longitudinal section of planula larva of *Gonothyraea*. Animal pole on top. B. Longitudinal section ▶ and cross section of a hydropolyp. Gastric cavity without septa. Tentacles with a solid endodermal axis are common. C. Longitudinal section of a hydromedusa. Apomorphy: velum on the umbrella rim composed of ectoderm of the exumbrella and subumbrella. el = endodermal lamella. ma = manubrium. me = mesoglea. ra = radial canal. ri = ring canal. st = stomach. te = tentacle. ve = velum. (A Hyman 1940; B Werner 1993; C after Thomas and Edwards 1991)

Fig. 31. Obelia dichotoma (Thecata, Hydrozoa). Formation of the medusa by lateral budding on the polyp. A. Tip of a polyp branch with gonangium – a modified polyp used for reproduction. The normal hydrotheca (cuticle of the hydranth) is transformed to an elongated gonotheca with a tight opening. Inside is a polyp without tentacles or mouth. The medusae originate on this blastostyle. B. Medusa with gonads at the radial canals. The velum is reduced. bl = blastostyle. gn = gonangium. go = gonad. gt = gonotheca. ht = hydrotheca. ma = manubrium. me = medusa bud. (Werner 1993)

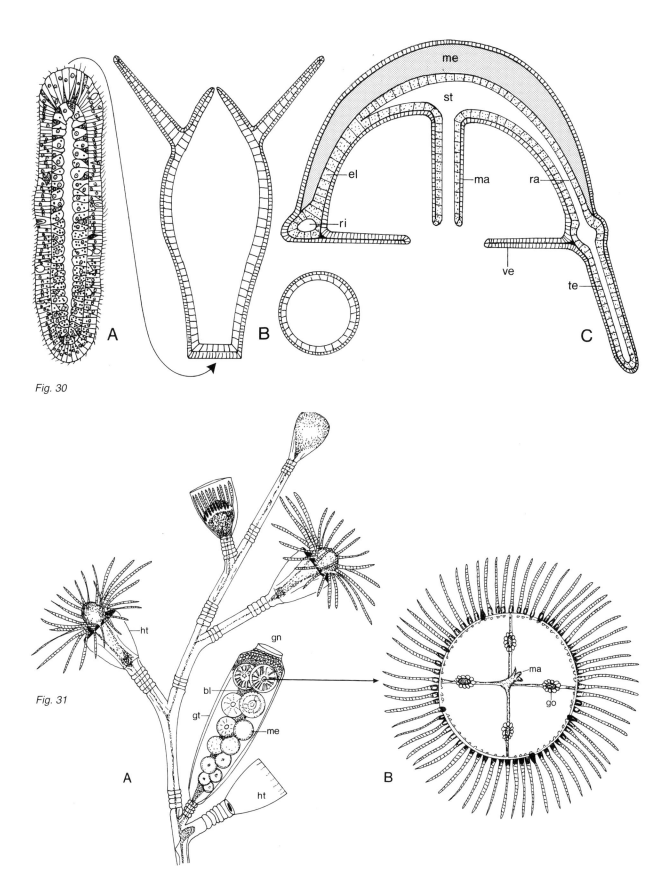

Fig. 30

Fig. 31

93

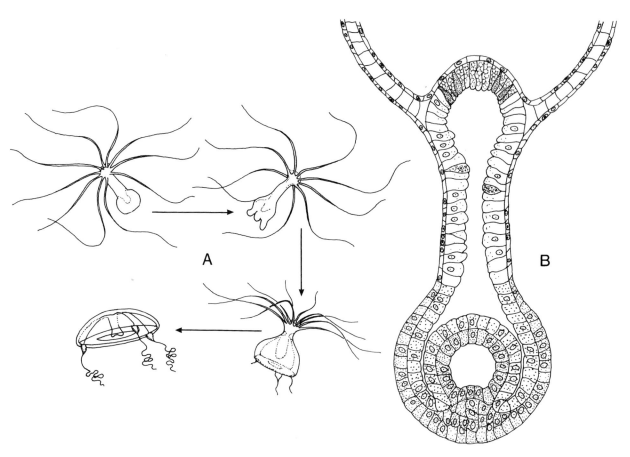

Fig. 32. Life cycle of *Eirene hexanemalis* (Narcomedusa, Hydrozoa) with metamorphosis of the polyp to medusa. A. The transformation takes place in open waters. The planula larva develops from the fertilized egg cell of a medusa; it develops into a pelagic, solitary polyp 0.5 mm in length. The ampulla-like lower section becomes a medusal bud. This develops into a medusa (2 cm in diameter), while the rest of the polyp, which carries the tentacle, degenerates. B. Longitudinal section of a polyp at the beginning of the transformation to medusa. The cavity in the ampulla is the subumbrellar cavity of the medusa. (Bouillon 1983)

Caution would seem necessary in regards to two further features which have been interpreted as autapomorphies (SCHUCHERT 1993). The first of these is the endodermal ring canal in the periphery of the hydromedusa. A comparable canal exists in the cubomedusa. If, however, the medusae of the Hydrozoa and Rhopaliophora are not homologous, then this similarity must be interpreted as a case of convergence.

The storage of gametes in the ectoderm, often referred to as a unique feature, is not uniform among the Hydrozoa. In addition, the development of gametes in the endoderm has been observed (TARDENT 1977).

Hollow tentacles are postulated as a plesiomorphy for the ground pattern of the Hydrozoa.

The question as to the original method by which medusae are produced by the polyp has not yet been resolved. The customary textbook presentation of a "typical", lateral bud from a sessile polyp (Fig. 31, 33) cannot be postulated as a part of the ground pattern of the Hydrozoa (SCHUCHERT 1993). In some Narcomedusa, a pelagic polyp, after releasing medusa buds, transforms

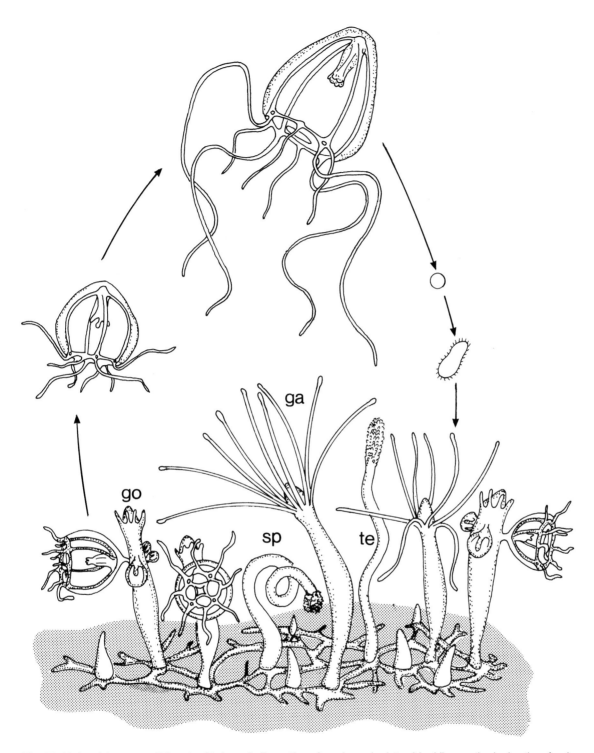

Fig. 33. Hydractinia carnea (Athecata, Hydrozoa). Formation of medusae by lateral budding on the hydranths of only slightly modified polyps. Also an example for polyp polymorphism in a hydrozoan colony. Distribution of elementary functions on differently differentiated polyps – gastrozooid (nourishment), tentaculozooid (prey-catching), spiralzooid (defense) and gonozooid (asexual reproduction). A stolon net is spread over the substrate which connects the polyps with each other. ga = gastrozooid. go = gonozooid. sp = spiralzooid. te = tentaculozooid. (Möhn 1984)

itself into a medusa (Fig. 32). This procedure is comparable to the complete metamorphosis of the polyp of the Cubozoa into a medusa, which we interpret to be a plesiomorphous characteristic within the Rhopaliophora.

A conclusive phylogenetic systematization of the Hydrozoa suitable for textbook publication does not yet exist. Several promising studies in this field are, however, already available (BOUILLON 1985, 1987; PETERSEN 1990).

Rhopaliophora

■ **Autapomorphies**
(Fig. 26 → 4)

- Medusa with rhopalia on the rim of its bell.

- Medusa with gastric filaments in the stomach.

- Tentacles of the polyp have a solid endodermal axis.

Rhopalia are complex, club-shaped sensory organs on the rim of the medusal bell. They number four in the Cubozoa and eight in the Scyphozoa. The individual sense organ is made up of ectoderm and endoderm. Endodermal cells produce a statolith composed of numerous calcium concretions. In both the Cubozoa and Scyphozoa, the statolith is made up of $CaSO_4$ (Fig. 34).

These two taxa are the only taxa of the Cnidaria to have four groups of gastric filaments in the stomach of the medusa in common. The finger-shaped outgrowths of the endoderm seize objects of prey; secretory cells secrete enzymes for the purpose of extracellular predigestion of food.

Finally, the polyps of the Cubozoa and Scyphozoa have, without exception, tentacles with a solid endodermal axis. This state is assessed by SCHUCHERT (1993) to be a further autapomorphy of the Rhopaliophora.

Cubozoa – Scyphozoa

We still have to justify one last high-ranking sister group relation within the Cnidaria. In order to do so, let us look at four prominent differences in alternative features.

The uniform coelenteron of the polyp is a plesiomorphy in the Cubozoa (Fig. 35), and the presence of four endodermal septa with septal funnels in the Scyphozoa is the apomorphy of the alternative (Fig. 37).

The Cubozoa exhibit a plesiomorphous total metamorphosis of the polyp into the medusa (Fig. 36); the reproductive behaviour of the polyps of the Scyphozoa by means of strobilation (terminal transverse fission) is, however, a derived feature.

Highly differentiated lens eyes in the Cubozoa are, in contrast, an apomorphy; comparable structures are absent in the Scyphozoa.

The velarium is a further characteristic feature of the Cubozoa which does not have an equivalent in the Scyphozoa.

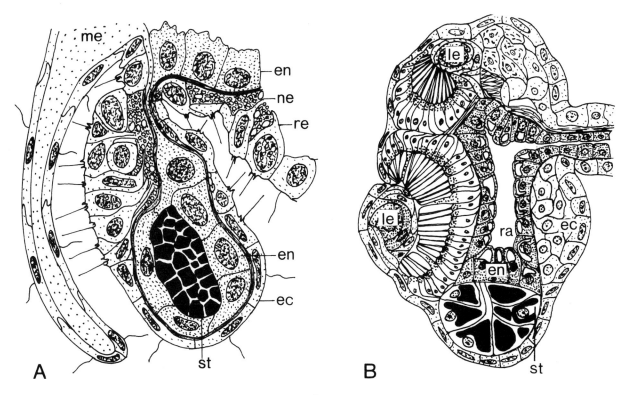

Fig. 34. Longitudinal section of rhopalia on the rim of the umbrella of scyphomedusa and cubomedusa. A. Rhopalium of *Nausithoe* (Scyphozoa). B. Rhopalium of *Tripedalia cystophora* (Cubozoa) with two lense eyes. The congruence in the construction of an endodermal statocyst with statoliths composed of $CaSO_4$ is assessed as a synapomorphy of the Scyphozoa and Cubozoa. ec = ectoderm. en = endoderm. le = lense eyes. me = mesoglea. ne = nerve fibers. ra = rhopalur ampulla. st = statocyst. (A Werner 1993; B Laska-Mehnert 1985)

Cubozoa

■ **Autapomorphies**
(Fig. 26 → 5)

– Lens eyes on the rhopalia.

– The velarium as an infolding of the subumbrella on the rim of the bell.

– Cube-shaped medusa.

– A nerve ring in the polyp.

The rhopalium of the cubomedusae has two median lens eyes next to the statocyst (Fig. 34) and two pit eyes on each side. WERNER (1993) regards rhopalia with this construction as the most complicated sensory organs of all the Cnidaria.

The following difference should be emphasized when comparing the velarium of the cubomedusae with the velum of the hydromedusae (p. 92); the velarium develops alone out of the subumbrella, i.e., out of the inner lining of the medusal bell; the velum, however, develops from the ecto-

97

dermal layers of the exumbrella and the subumbrella. The cubeform of the cubomedusae with an almost square cross section is unique within the Cnidaria.

A characteristic autapomorphy is the nerve ring of the cubopolyp. A closed ring of nerve fibers surrounds the mouth just above the base of the tentacles. This ring is composed of ectodermal and endodermal parts. Comparable formations are not known in the polyps of the other taxa of the Cnidaria.

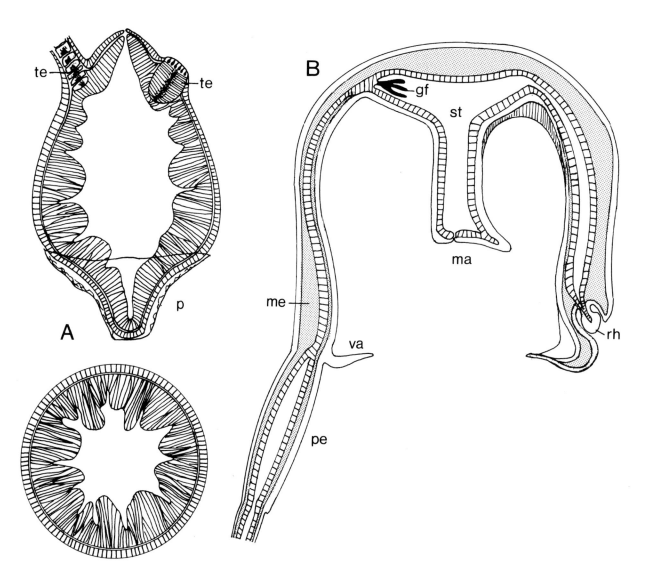

Fig. 35. Cubozoa. A. Longitudinal section and cross section of the cubopolyp of *Tripedalia cystophora*. B. Organization of the cubomedusa in a longitudinal section. gf = gastric filament. ma = manubrium. me = mesoglea. p = periderm cup. pe = pedalium. Leaf-like, spread-out basis of the tentacle. rh = rhopalium. st = stomach. te = tentacle. va = velarium. (Werner 1993)

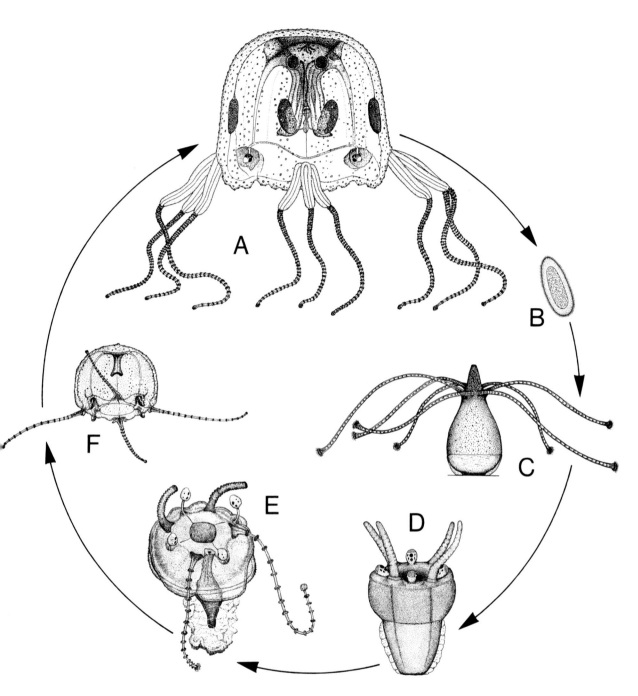

Fig. 36. Life cycle of *Tripedalia cystophora* (Cubozoa) with complete metamorphosis of the polyp to medusa. A. Medusa with gonads. B. Planula larva. C. Young sessile polyp. D. Beginning of the transformation with tentacle of the polyp changing into rhopalia and the development of the medusal tentacle. E. Final phase of metamorphosis. Only a rest of the polyp remains in the shrunken peridermal sheath. F. Young free-swimming medusa (diameter of umbrella 1.5 mm). (Werner, after Ax 1988)

Scyphozoa

The four endodermal septa give the scyphopolyp a distinct tetraradial symmetry (Fig. 37). In comparison with the septa of the Anthozoa, the diverging number of septa in the Scyphozoa and the basic difference in their musculature speak against a homology.

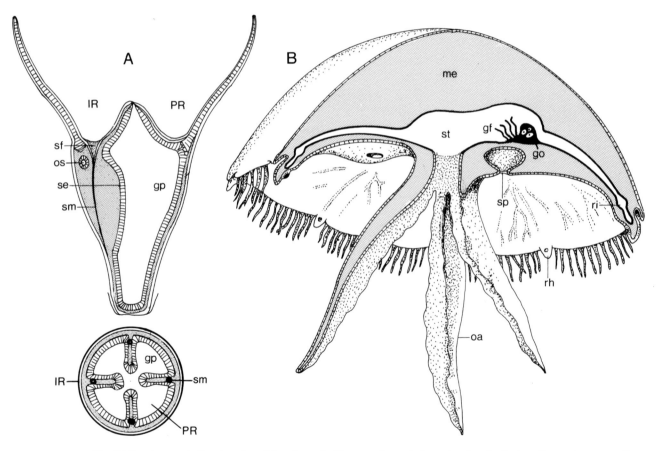

Fig. 37. Scyphozoa. A. Scyphopolyp of *Aurelia aurita* (Semaeostomea) shown in a longitudinal section and cross section. The longitudinal section runs left through an interradius (IR), right through a perradius (PR). Tentacle with a solid endodermal axis as found in the Cubozoa. Autapomorphy: four endodermal septa with four ectodermal funnels carrying longitudinal muscles. B. Diagram of a scyphomedusa with four long oral arms (Semaeostomea). Special attention should be paid to four groups of gastric filaments in the stomach (synapomorphy with the Cubozoa) and four ectodermal subgenital pits underneath the gonads (most likely an autapomorphy of the Scyphozoa). gf = gastric filament. go = gonad. gp = gastral pouch. me = mesoglea. oa = oral arm. os = septal ostium. rh = rhopalium. ri = ring vessel. se = septum. sf = septal funnel. sm = septal muscle. sp = subgenital pit. st = stomach. (A Werner 1993; B Bayer and Owre 1968)

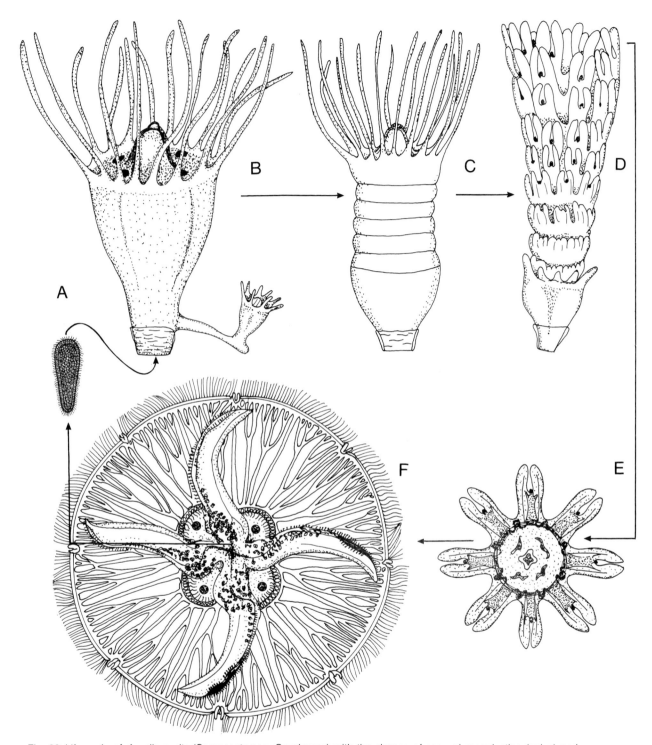

Fig. 38. Life cycle of *Aurelia aurita* (Semaeostomea, Scyphozoa) with the change of asexual reproduction (polyp) and sexual reproduction (medusa). A. The planula larva attaches itself with the apical pole. B. It develops into a sessile, solitary scyphostoma polyp (height 3 mm); it can produce new polyps through lateral buds. C. Beginning of strobilation (formation of medusae) through ring-shaped constrictions of the body of the polyp. D. Advanced strobilation with a set of young medusae which lie on top of each other but are still attached to the body of the polyp. E. The upper medusal bud has freed itself in the form of an ephyra (diameter 2 mm). F. The free-swimming ephyra has developed into an adult (width of umbrella 25 cm). The females carry the fertilized eggs in brood pouches of the oral arms, where development takes place up to the planula stage. (After Werner 1993)

The ectodermal funnels of the scyphopolyp sink deep into the individual septa from the buccal field near the body wall. Ectodermal myocytes differentiate themselves from these septal funnels and join together in the Mesoglea forming bundles of muscles cells.

Scyphopolyp contraction has a completely different structure and genesis against the contraction in the Anthozoa. Anthozoan muscle bands are made up of epithelial muscle cells, lie laterally to the septa, and are of endodermal origin (p. 89).

The unique process of strobilation of the Scyphozoa by the budding of ephyrae with eight lappets (Fig. 38) can be derived from the total metamorphosis of the polyp as seen in the Cubozoa and as

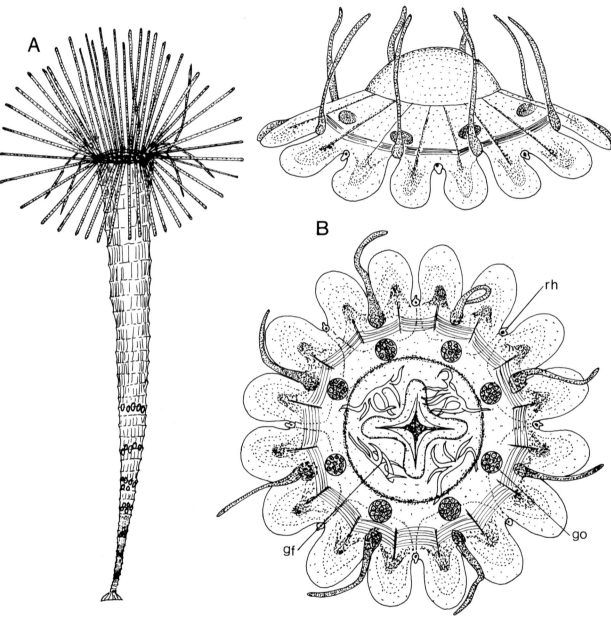

Fig. 39. Polyp and medusa of Coronata (Scyphozoa). A. *Stephanoscyphus*. The polyp is surrounded by a firm, horn-shaped cuticula (periderm). B. Side view of medusa of *Nausithoe* (1–2 cm in diameter) and view of the oral side. The medusa of the Coronata matures during the ephyral stage. A derived feature is the circular cleavage which divides the body of the medusa into the central section and periphery. gf = gastral filament. go = gonad. rh = rhopalium. (Werner 1993)

it belongs to the ground pattern of the Rhopaliophora. In the stem lineage of the Scyphozoa, a partial rearrangement and transversal fission of the upper part of the polyp evolved from the complete metamorphosis of the polyp into the medusa. Monodisk strobilation, the fission of a single medusa (such as occurs in the Rhizostomea) may serve as a model for the evolutionary path to polydisk strobilation, which is characterized by the simultaneous formation of several disk-shaped ephyrae.

The subtaxa **Stauromedusida**, **Coronata**, **Semaeostomea**, and **Rhizostomea** (Figs. 38-40), conventionally known as orders, appear to represent monophyla. Their phylogenetic relations to each other have not yet been adequately clarified.

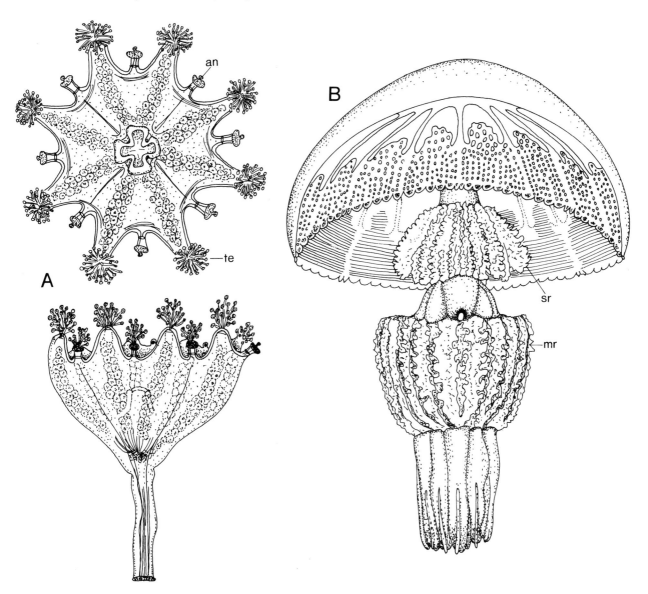

Fig. 40. Scyphozoa. A. *Haliclystus salpinx* (Stauromedusida) with a view of the oral disk and a side view. A noticeable autapomorphy is the absence of metagenesis. A stem-shaped section of the polyp, which widens at the top to form a cup-shaped medusal section, grows out from the planula. Eight adhesive organs (marginal anchors) are positioned between the eight bundles of short, buttoned tentacles. B. *Rhizostoma octopus* (Rhizostomea). The primary mouth is lost due to fusion of the oral lips. Numerous secondary openings are found in the mouth and shoulder frill areas. an = anchor organ. mr = mouth ruffles. sr = shoulder ruffles. te = tentacle. (Werner 1993)

103

Acrosomata

■ **Autapomorphies**
(Fig. 25 → 2)

– Sperm with uniform acrosome and a subacrosomal perforatorium.

– Real myocytes between ectodermal cells and below the epidermis embedded in ECM.

The two features interpreted to be synapomorphies of the Ctenophora and Bilateria have already been discussed (p. 82).
We will now go on to justify the adelphotaxa relation between the Ctenophora and the Bilateria.

Ctenophora – Bilateria

A benthic organism without axes of symmetry has been hypothesized as the stem species of the Metazoa (p. 58). Correspondingly, a soil-dwelling stem species without symmetry and with nondirectional locomotion must be postulated for the Eumetazoa and the Acrosomata.
Starting from this condition, then the Ctenophora and the Bilateria have followed different evolutionary paths. The move from the benthal to the pelagial took place in the stem lineage of the Ctenophora. Biradial symmetry (disymmetry) and a number of other characteristics evolved during this shift. In contrast, the phylogenesis of the Bilateria took place in the primary biotope, the benthal. The evolution of unidirectional locomotion on the ocean floor triggered the formation of a bilaterally symmetrical body and of numerous new organizational traits.

Ctenophora

■ **Autapomorphies**
(Fig. 25 → 3)

– Medusal organism of the pelagial with extensive mesoglea.

– ? Two tentacles with collocytes (adhesive cells).

– Biradial symmetry with sagittal and transverse planes, each of which divides the body into symmetrically equal halves.

– Eight comb rows which are the primary means of locomotion.

– Statocyst and two polar fields at the aboral pole.

– Gastrovascular system has an axial and a peripheral section (Figs. 41, 42).
A long, laterally compressed ectodermal pharynx is attached to the mouth opening in the axis of the body. It is followed by the endodermal stomach, from which the aboral canal rises vertically. The latter divides into four branches, two of which extend outwards and are called anal pores.

The eight meridional canals beneath the comb rows and their horizontal connections with the stomach are part of the ground pattern of the Ctenophora. This is less certain in the case of the two tentacle vessels and the pharyngeal vessels, which are sometimes absent.

– Hermaphroditism.

– Position of the gonads in the wall of the meridional canals.

– Biradial cleavage.

– Mosaic development.

– Movement with the oral pole first.

Ground Pattern

A habitus with a round cross section, as is found in *Pleurobrachia pileus* of the North and Baltic Seas, is usually assumed to be plesiomorphous. Two tentacles are also considered to form part of the ground pattern of the ctenophores. They arise from the ectodermal tentacle pouches and possess characteristic collocytes (adhesive cells, see Fig. 41).

On the other hand, the majority of sharply deviating ctenophores including the ribbon-shaped Cestida (*Cestus, Velamen*) and the flattened Platyctenida (*Ctenoplana, Coeloplana*) in the benthal can be considered to be derived, since their ontogenesis includes a spherical cydippid larva. There is even a complete reduction of tentacles within the Lobata. *Ocyropsis maculata* has rudimentary tentacles; the adult *Ocyropsis crystallina* has no tentacles at all. These species catch their prey by means of oral lobes (HARBISON 1985).

The Beroida (*Beroe, Neis*), which have no tentacles and which are shaped like a flat cap or miter, are active predators (Fig. 42). They do not even have tentacles during their development. For this reason, HARBISON (1985) contrasts the hypothesis of a stem species without tentacles and with a predatory mode of nourishment with the hypothesis that the ground pattern of the Ctenophora features tentacles and a passive means of acquiring food through collocytes. Since there is nothing comparable to the colloblasts in the field surrounding the Ctenophora, the alternative in the out-group comparison cannot be determined. One could support a secondary absence of tentacles in the Beroida by arguing that there is evidence for an evolutionary path of this kind within the Lobata, even if the type of predation is different.

Tentacles with collocytes (colloblasts) have been marked with a question mark in the list of autapomorphies, because they may not be part of the ground pattern of the ctenophores. Instead, they may have evolved in the stem lineage of a subunity Tentaculifera, comprising all Ctenophora except for the Beroida.

In any case, a single evolution can be postulated for the **collocytes** (Fig. 43), which are products of the differentiation of individual ectodermal cells. The head of the mushroom-shaped adhesive cell protrudes over the surface of the epidermis. The secretory granules are arranged peripherally in the head and cause the adhesion of objects of prey. Radial fibers anchor the granula with a spheroidal body in the center of the cell head. An elongated nucleus is attached to it. The elastic element of the collocyte is the helical thread, which is a highly modified cilium. The ciliary rootlet is

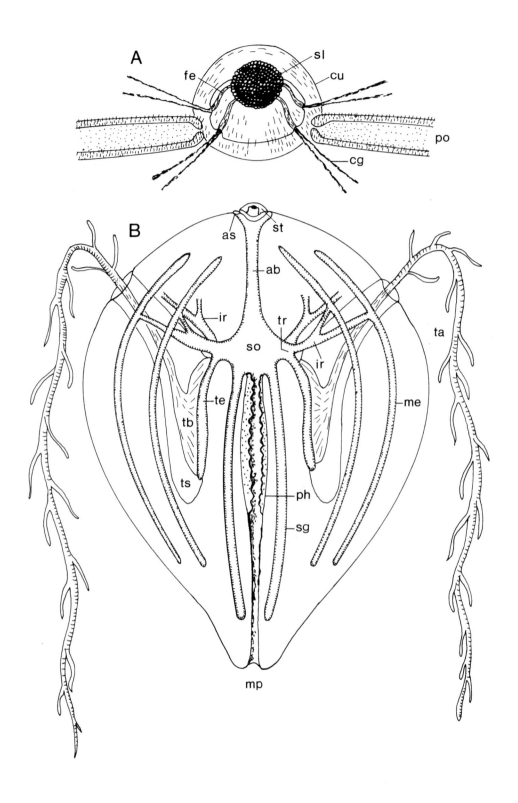

Fig. 41. Ctenophora. A. Aboral sensory pole with statocyst and polar fields. B. Diagram of the probable ground pattern. ab = aboral canal. as = anal pore. cg = ciliary pit. cr = comb row. cu = cup of cilia. fe = feathers. ir = interradial canal. me = meridional canal. mp = mouth. ph = pharynx. po = polar field. sg = pharyngeal canal. sl = statolith. so = stomach. st = statocyst. ta = tentacle. tb = tentacle basis. te = tentacle vessel. tr = transversal canal. ts = tentacle sack. (Hyman 1940; after Hernandez-Nicaise 1991)

106

located in the base of the cell. The thread winds several times around the adhesive cell and ends on the spheroidal body. The helical thread is surrounded by plasma membranes of the collocyte and is connected to it by means of a thin ridge.

The development of an extensive **mesoglea** (ECM) with collagen fibrils correlates to the evolution of a medusoidal organism of the macroplankton. We must assume that this mesoglea originated in the stem lineage of the Ctenophora convergently to the homonymous mesoglea of the Cnidaria.

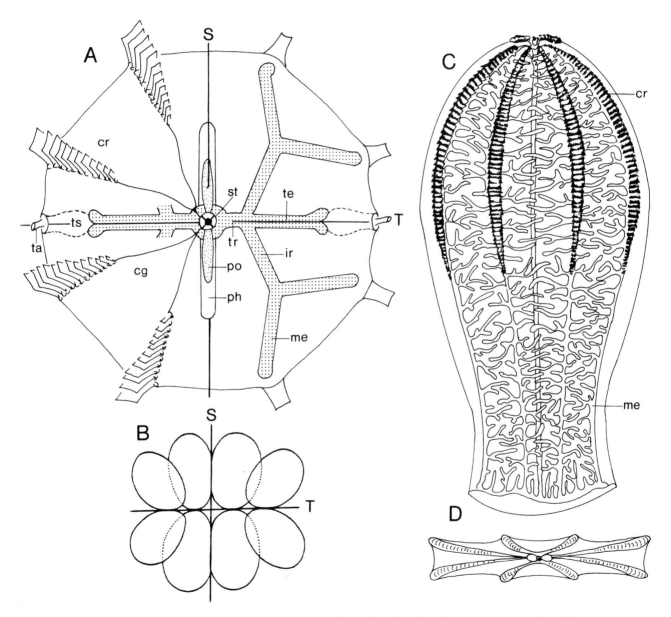

Fig. 42. Ctenophora. A and B. Demonstration of biradial symmetry. Dissection of the body through the sagittal plane (S) and the transversal plane (T) into symmetrical halves. A. View of the apical pole in the probable state of the ground pattern. B. Early stage of cleavage of *Beroe ovata* with eight blastomeres arranged on one plate. C. *Beroe* with cap-shaped habitus. D. *Beroe.* View of the aboral pole. The body is greatly flattened in the transversal plane. Explanation of abbreviations in Fig. 41. (A Hernandez-Nicaise 1991; B-D Hyman 1940)

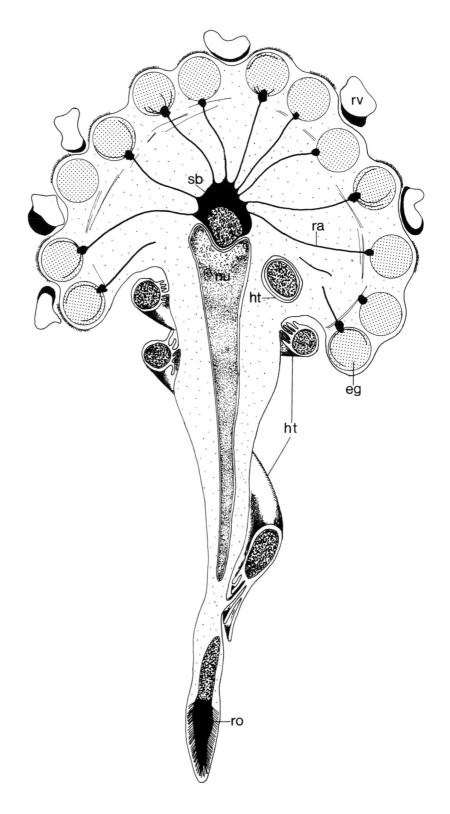

Fig. 43. Ctenophora. Ultrastructure of an adhesive cell (collocyte) in *Pleurobrachia*. eg = esosinophile granules. ht = helical thread. nu = nucleus. ra = radial fibrils. ro = rootlet. rv = refractory vesicle. sb = spheroidal body. (Hernandez-Nicaise 1991)

The **musculature** of the Ctenophora was adopted from the ground pattern of the Acrosomata in the form of individualized myocytes. The muscle cells in the ectoderm and endoderm form a special parietal musculature. Bundles of syncytial muscle fibers are embedded in the mesoglea.

In the **biradial symmetry** or disymmetry of the ctenophores, two planes of symmetry meet at right angles to each other at the vertical main axis, which runs from the apical sensory pole to the oral opening. The body can be dissected into two symmetrical halves along the sagittal plane (pharyngeal plane) and the transverse plane (tentacular plane; Fig. 42).

The ectodermal pharynx is flattened in the sagittal plane; the two ciliated polar fields of the apical pole are located in this plane. The stomach is compressed in the transversal plane. The two tentacles, when present, are located in this plane. It should be mentioned at this point once more that the biradial symmetry of the planktonic Ctenophora has nothing to do with the biradial symmetry in single taxa of the sessile Anthozoa (p. 89).

Let us now discuss the **eight comb rows** and their arrangement in the four squares of the body as another characteristic, apomorphous feature of the Ctenophora (Fig. 42). Basic elements of locomotion are the comb plates of the rows on cushion-like elevations of the epidermis. Here, the extremely long and thin epidermal cells each form 15–50 cilia. A single comb plate can have 100 000 cilia (HERNANDEZ-NICAISE 1991). The existence of multiciliated cells is of importance in the comparison with the sister group Bilateria.

Monociliated epidermal cells, such as found in the Gnathostomulida and a part of the Gastrotricha Macrodasyoida, belong to the ground pattern of the Bilateria. Accordingly, the multiciliation in the comb plates of the ctenophores can be conclusively interpreted as a convergence to the widespread multiciliated epidermis within the Bilateria.

The **statocyst** on the aboral pole of the Ctenophora (Fig. 41) is also an autapomorphy of the Ctenophora. It is not homologous with the statocysts in the medusae of the Cnidaria, or with the static sensory organs within the Plathelminthes. The structure of the statocyst – with a statolith made of calcium sulfate which is carried by four balancers of cilia and is covered by a dome of other cilia – is not seen in other animals.

The **gastrovascular system** has already been outlined in the list of autapomorphies. The entire structure, with the pharynx running deep into the body and the eight meridional canals, is unique and cannot be compared with the medusae of the Cnidaria. One plesiomorphous congruence should, however, be mentioned for the sake of completeness. This is the common existence of a single opening originating from the embryonic blastopore region for the intake of food and the release of excrements.

We postulate gonochorism for the stem species of the Metazoa and the Bilateria. The **hermaphroditism** of the Ctenophora is thus seen as an autapomorphy. Both the *Ocyropsis* species mentioned above were recently described as probably being gonochoric Ctenophora (HARBISON and MILLER 1986); within the Ctenophora they are, however, referred to as the descendants of a stem species with simultaneous hermaphroditism.

The position of the gonads on the meridional canals remains a characteristic feature of the Ctenophora. However, its correlative link to the construction of the gastrovascular system makes this less significant. The production of gametes in the endoderm is a plesiomorphy.

In contrast to radial cleavage and the ability of embryonal regulation in the Cnidaria and within the Bilateria, the biradial (disymmetrical) cleavage and mosaic development in the **ontogenesis** of the Ctenophora are clearly apomorphous features. The first three cleavages run meridionally and produce a disk made up of eight blastomeres (Fig. 42); the biradial symmetry of the Ctenophora appears at an early stage due to this apomorphous pattern of cleavage. On the other hand, direct development, without stages of ontogenesis with specific larval features, is a plesiomorphy in the ground pattern of the Ctenophora.

A last fundamental difference between the Ctenophora and the cnidarian medusae is their diametrically opposed mode of locomotion. Cnidarian medusae swim with the aboral pole to the fore, whereas the Ctenophora swim with the oral pole in front as a result of the aborally directed strokes of the comb plates.

Bilateria

■ **Autapomorphies**
(Fig. 25 → 4)

– Polar organization of the body with bilateral symmetry. A sagittal plane divides the body in the longitudinal axis into right and left halves.

– Basal lamina as a condensation of the ECM beneath the ectoderm and endoderm into a strongly contrasting layer.

– Body wall musculature made up of outer circular muscles and inner longitudinal muscles.

– Paired protonephridia of ectodermal origin each consisting of three monociliated cells: a terminal cell, canal cell, and nephropore cell. Terminal cell with eight microvilli around the cilium and with a distal porous cylinder, which is the supporting structure for the ECM filter.

– Central nervous system with brain at the front end.

Ground Pattern

A series of new organizational traits appear in the Bilateria as against the Placozoa, Cnidaria, and Ctenophora. A detailed representation of the ground pattern is, for this reason, necessary. Each feature should be examined to determine what was adopted into the ground pattern as a plesiomorphy, and what developed in the stem lineage of the Bilateria as an evolutionary novelty (apomorphy).

Body Size, Habitus, and Symmetry
The body of *Trichoplax adhaerens* (Placozoa) and of numerous species of diverse Bilateria taxa (Gnathostomulida, the majority of the free-living Plathelminthes, Gastrotricha, free-living Nematoda, and others) is only a few millimeters in size. There are no reasons in the case of any of the taxa mentioned to consider this size to be a derived state. Consequently, we postulate an organism for the stem species of the Bilateria measuring some few millimeters, and consider this to be a plesiomorphy.

A bilaterally symmetrical body evolved in the stem lineage of the Bilateria from an asymmetrical organism, as is represented by *Trichoplax adhaerens*. In the ground pattern of the Bilateria, the sagittal plane divides the body into right and left symmetrical halves along a longitudinal axis from the front end to the back end of the body (Fig. 44). The transversal or horizontal plane divides the organism into two unequal halves, a dorsal and ventral section. **Bilateral symmetry** is the first essential autapomorphy of the Bilateria.

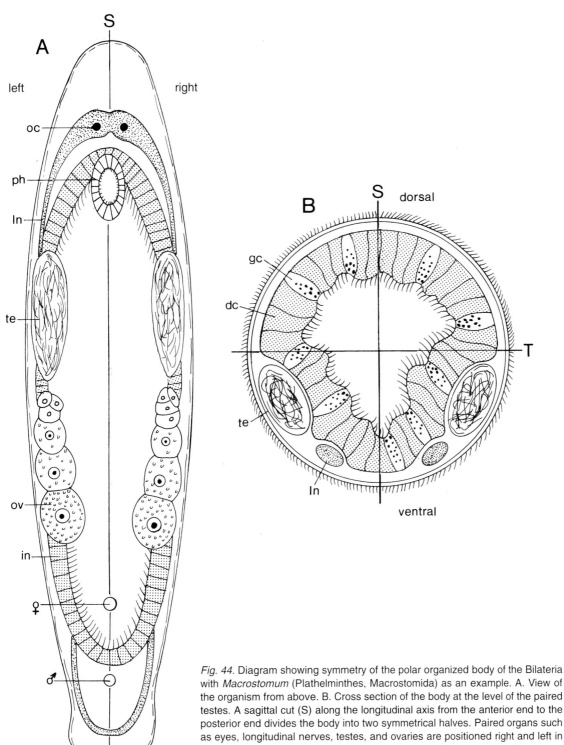

Fig. 44. Diagram showing symmetry of the polar organized body of the Bilateria with *Macrostomum* (Plathelminthes, Macrostomida) as an example. A. View of the organism from above. B. Cross section of the body at the level of the paired testes. A sagittal cut (S) along the longitudinal axis from the anterior end to the posterior end divides the body into two symmetrical halves. Paired organs such as eyes, longitudinal nerves, testes, and ovaries are positioned right and left in corresponding places. Unpaired formations such as the pharynx and the genital openings are often arranged in the sagittal plane. The transversal plane (T) divides the body into a dorsal and a ventral half which are, as a rule, asymmetrical. In *Macrostomum*, the brain and eyes are located in the upper half whereas the testes, ovaries, and longitudinal nerves are located in the lower half. dc = digestive cells. gc = glandular cells. in = intestine. ln = longitudinal nerve. oc = ocellus. ov = ovary. ph = pharynx. te = testis.

A habitus with a round cross section and with gradually tapered front and back parts is evidently related to the evolution of bilateral symmetry. A body structure of this kind is not only seen in the Gnathostomulida or the Nematoda; contrary to popular belief, and despite the associations evoked by the term "flatworm", the overwhelming majority of free-living Plathelminthes also have a perfectly round body without a trace of dorsoventral compression.

Biotope and Locomotion
The Bilateria are primarily inhabitants of surface sediments of the benthal. The change from the nondirectional locomotion of an organism without symmetry to unidirectional locomotion on the ground, with the formation of an anterior and posterior pole, probably triggered the evolution of bilateral symmetry. This then led to the concentration of sensory equipment and to the origin of a brain-like center of the nervous system in the anterior end.

Epidermis
Like the Placozoa, various unities of the Bilateria have monociliated ectodermal cells with a diplosomal basal apparatus. In the Gnathostomulida, the body is completely equipped with monociliated cells. The ventral ciliation in a number of Gastrotricha species is with monociliated epidermal cells. In addition, they were described in larvae of the Radialia ("Tentaculata", Hemichordata, Echinodermata; overview: RIEGER 1976). The epidermal pattern with one cilium per cell is doubtless a plesiomorphy of the Bilateria, and is a part of the ground pattern of the unity.

Intestine
The lack of an anus in the Gnasthostomulida and Plathelminthes must be interpreted as a plesiomorphy common to the Cnidaria and Ctenophora; there are no arguments to substantiate a secondary reduction. The intestine without an anus is linked with the ventral oral opening at the anterior end. This is a standard feature in the Gnathostomulida, and common among the Plathelminthes. For this reason, it seems legitimate to postulate both of these features in the ground pattern of the Bilateria.

The intestinal cells in many of the unities of the Bilateria are nonciliated. Monociliated gastrodermal cells appear in the Radialia (Brachiopoda, Acrania; RIEGER 1976). On the other hand, multiciliated endodermal cells are part of the ground pattern of the Plathelminthes, the Nemertini, and the Rotatoria. The hypothesis that monociliated intestinal cells were adopted from the level of the Placozoa into the ground pattern of the Bilateria remains, however, the most parsimonious explanation.

Basal Lamina
In numerous Bilateria, the ECM is condensed into a strongly contrasting layer beneath the ectoderm and endoderm. This basal lamina or basal membrane is assessed by BARTOLOMAEUS (1993a) to be an autapomorphy in the ground pattern of the Bilateria.

Mesoderm and Body Cavity
The terms mesoderm and body cavity are closely linked to one another in the Bilateria. After defining the mesoderm as a tissue in the extracellular matrix between the ectoderm and endoderm (p. 81), it is now necessary to describe the different manifestations of this body layer and to characterize the state which may be postulated for the ground pattern of the Bilateria. As regards the body cavities, we will first concentrate on extensive cavities which function as hydroskeletons. Cavities which are proven to be paths of transport for blood plasma will be discussed in the next section.

1. Compact (acoelomate) Organization Without Cavities.
(Fig. 45A)

a) Ectoderm and endoderm are closely joined together.

Myocytes of the body musculature are embedded in the ECM in the gaps between the epithelia. Various other cells, such as sensory cells, nerve cells, and glandular cells are also embedded there. Stem cells (neoblasts) are found in the Plathelminthes. In addition, the sex organs are located in the compact middle layer.
This state characterizes the Gnathostomulida, the majority of the millimeter-sized Plathelminthes, the Gastrotricha, and the free-living, small Nematoda.

b) Ectoderm and endoderm shift farther apart.

Larger gaps between the epithelia and the reproductive organs are filled by parenchymatous cells. This condition appears in large, mostly flat Plathelminthes. Well-known textbook examples include the free-living Tricladida (planarians) and parasites such as the *Fasciola hepatica* (Trematoda).

The described organization leads us first to the following conclusion about the Plathelminthes. In the correlation to a primarily small body, the compact organization with narrow gaps between the ectoderm and endoderm is a feature of the ground pattern of the Plathelminthes. Voluminous parenchymatous bodies originated only secondarily in connection with the increase in body size from millimeters to centimeters within the Plathelminthes. This occurred repeatedly, and convergently, in the free-living and parasitic flatworms. Authors of textbooks should therefore refrain from describing cross sections with extensive parenchyma (mesenchyme) as "typical" for the Plathelminthes. Reference to the space between the ectoderm and endoderm as "schizocoel" should also be avoided, and the Plathelminthes should not be referred to as "Parenchymia" (EHLERS 1995).
The terms "compact organization" or "compact state" (RUPPERT 1991) should be used in preference to the popular term acoelomate. Acoelomate (= without a coelom) signifies only the alternative to coelomate (= with a coelom), and therefore does not include the primary body cavity, which we will now characterize in the following section.

2. Primary Body Cavity
(Fig. 45B)

"Cavity between ectoderm and endoderm enclosed by the extracellular matrix" (BARTOLOMAEUS 1993a).

Let us again take a familiar example often used in textbooks – the body cavity of large, parasitic nematodes. Many young academics have had unpleasant experiences with the fluid contents of the extensive body cavities of the *Ascaris suum*, whose body is under high hydrostatic pressure. This body cavity features no traces of an epithelial lining.
A simple comparison within the monophylum Nematoda quickly shows that, among the parasites, this is an extremely derived state of the body cavity (Fig. 46). Free-living nematodes have a compact organization without any cavity (BARTOLOMAEUS 1993a; EHLERS 1994b).

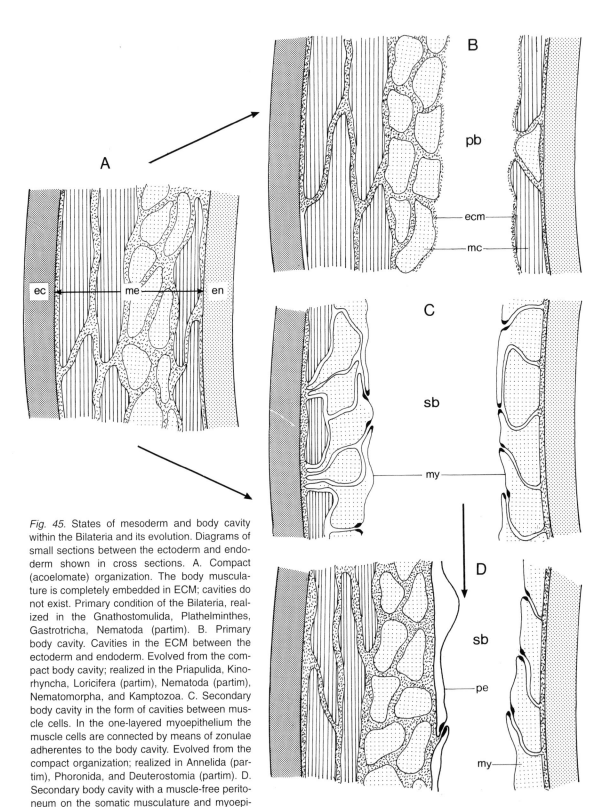

Fig. 45. States of mesoderm and body cavity within the Bilateria and its evolution. Diagrams of small sections between the ectoderm and endoderm shown in cross sections. A. Compact (acoelomate) organization. The body musculature is completely embedded in ECM; cavities do not exist. Primary condition of the Bilateria, realized in the Gnathostomulida, Plathelminthes, Gastrotricha, Nematoda (partim). B. Primary body cavity. Cavities in the ECM between the ectoderm and endoderm. Evolved from the compact body cavity; realized in the Priapulida, Kinorhyncha, Loricifera (partim), Nematoda (partim), Nematomorpha, and Kamptozoa. C. Secondary body cavity in the form of cavities between muscle cells. In the one-layered myoepithelium the muscle cells are connected by means of zonulae adherentes to the body cavity. Evolved from the compact organization; realized in Annelida (partim), Phoronida, and Deuterostomia (partim). D. Secondary body cavity with a muscle-free peritoneum on the somatic musculature and myoepithelial cells around the intestine. Evolved from the purely muscle-lined secondary body cavity (condition C); realized in the Annelida (partim), Echiurida, Sipunculida and Echinodermata (partim). ec = ectoderm. ecm = extracellular matrix (ECM). en = endoderm. mc = muscle cell. me = mesoderm. my = moepithelial cell. pb = primary body cavity. pe = muscle-free peritoneal cell. sb = secondary body cavity. (After Bartolomaeus 1994)

114

The primary body cavity is the result of an evolutionary increase in body size within the Nematoda, analogous to the evolution of a voluminous, parenchymatous body within the Plathelminthes.

A primary body cavity is not part of the ground pattern of the Nematoda, and obviously therefore does not belong to the ground pattern of a taxon Nemathelminthes as far as this can be justified as a monophylum (p. 130).

Consequently, the term pseudocoel is unsuitable for the Nematoda (EHLERS 1994b). It is simply wrong to use this term to refer to a supposed alternative between the "body cavity" of the Nemathelminthes (or even only the Nematoda) and the coelom of the Annelida. The word pseudocoel is superfluous and should not be used in zoology textbooks.

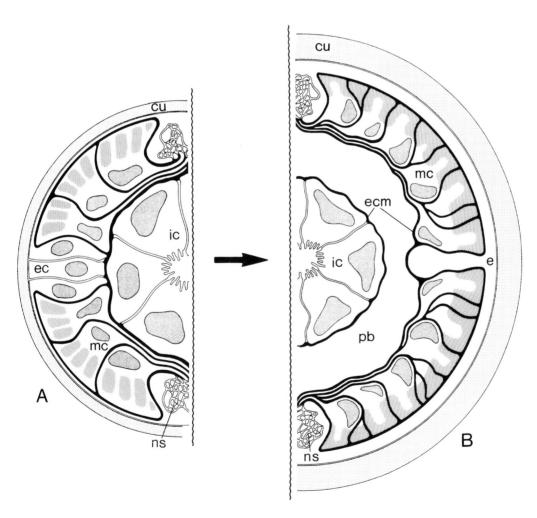

Fig. 46. Nematoda. A. Compact (acoelomate) organization as feature of the ground pattern of the Nematoda. The extracellular matrix (ECM = thick black lines) forms a continuum between the epidermis and intestine; cells of the mesoderm (muscle cells, gametes) are embedded in the ECM. This plesiomorphous state is realized in small, free-living Nematoda. B. Primary body cavity in the form of enlarged gaps in the ECM, surrounding the intestine as cavity. Apomorphous state; realized in large parasitic Nematoda. cu = cuticula. e = epidermis. ec = epidermis cell. ecm = extracellular matrix. ic = intestinal cell. mc = muscle cell. ns = nervous system. pb = primary body cavity. (Bartolomaeus 1993a)

3. Secondary Body Cavity
(Fig. 45C,D)

"Cavity lined with cells between the ectoderm and endoderm."

Let us concentrate on the extensive cavities between the ectoderm and endoderm which function as a hydroskeleton. Secondary body cavities of this nature are developed within the Spiralia[5] in the Sipunculida, Echiurida, the polymerous Annelida, and in the taxon Radialia[5] in the oligomerous "Tentaculata", Hemichordata, and Echinodermata.

The terms secondary body cavity and coelom are synonymous amongst these unities. This, however, says nothing about the homology or the nonhomology between the cavities themselves, nor about the existence of a taxon Coelomata that comprises the groups having a secondary body cavity.

To begin with, it should be noted that the conventional opinion that the coelom cavities in question are completely lined by a nonmuscular coelom epithelium is in need of revision. A peritoneum of this nature does not exist at all. The primary form of the lining of the body cavity is a one-layered myoepithelium made up of epithelial muscle cells (RIEGER 1986; RIEGER and LOMBARDI 1987; FRANSEN 1988; BARTOLOMAEUS 1993a,1994).

Nonmuscular peritoneal cells appear in the coelom of the Echinodermata. They also appear, though only laterally in the region of the somatic muscles, in the Annelida, Sipunculida and Echiurida. They always lie on the inside of the layers of muscle cells, i.e., they point towards the cavity of the coelom. A partially nonmuscular peritoneum develops secondarily during ontogenesis from the division and transdifferentiation of myocytes which are the first "coelom cells" to appear.

This brings us to the development of the coelom lining in the ontogenesis of the Annelida. Coelomogenesis takes place postlarvally in a zone of differentiation in front of the pygidium. The muscle cells of the prospective skin and intestinal muscle system always develop first from two caudolateral tissue cones. The differentiation of myocytes precedes all other cells which could partake in the lining of the coelom (peritoneal cells, podocytes). From the very beginning, the muscle cells are forming an epithelium with the following polarity. The cell bases position themselves on the ECM underneath the skin and on the intestine. They connect with each other apically by means of zonulae adherentes towards the body cavity.

During the course of coelomogenesis, gaps develop between the somatic and visceral epithelial muscle layers. The gaps fuse together and extend to become the cavity of the coelom (BARTOLOMAEUS 1993a).

The following facts and the logical conclusions to be drawn from them would seem relevant to our assessment.

a) The peripheral body musculature and the subepithelial intestinal musculature of the Annelida form at the same time the somatic and visceral lining of the secondary body cavity in the form of myoepithelia. In other words, the same tissue plays a vital role during locomotion, and functions as an abutment for the hydrostatic pressure of the coelom fluid.

b) The coelom wall of the Annelida develops exclusively from myocytes. During coelomogenesis, the muscle cells differentiate in front of the pygidium from a compact embryonal tissue between the ectoderm and endoderm.

---

5 Justification of the unities as monophyla on p. 130.

c) These facts suggest the evolutionary derivation of the secondary body cavity of the Annelida from the compact state of primarily acoelomate Bilateria. Muscle cells are realized here in the narrow gaps between ectoderm and endoderm only in the form of pure myocytes in the ECM. The hydrostatic pressure resulting from the evolution of fluid filled cavities in the muscle system requires the development of an abutment which manifests itself in the epithelial arrangement of the myocytes with apical cell connections.

d) The muscle cells of the Annelida all have a pair of centrioles which occasionally induce the formation of rudimentary cilia (GARDINER and RIEGER 1980; BARTOLOMAEUS 1993a, 1994). This plesiomorphous feature – muscle cells with a diplosome – originates directly from the compact organization of the Bilateria. Here the myocytes of the Gnathostomulida are all equipped with a pair of centrioles (LAMMERT 1991), as are muscle cells of the Nemertini (BARTOLOMAEUS 1994). Ciliogenesis is obviously not possible when the muscle cells are embedded in a compact organization, but is possible when the muscle cells border a coelom cavity such as in the Annelida and Echinodermata.

Conclusion

Our analysis leads us to propose a compact acoelomate organization for the ground pattern of the Bilateria. Repeatedly, convergent cavities evolved as hydroskeletons from this plesiomorphous state, both as primary body cavities within the Nemathelminthes, and as secondary body cavities with a myoepithelial lining within the Spiralia and in the stem lineage of the Radialia. The possibility of a general homology of the cavities, referred to as coelom in the Bilateria, is to be rejected. Unfortunately, a traditional term is used for different nonhomologous organizational traits of the Bilateria. Further reasons for this position are based on the analysis of the nephridial organs which follows below.

In the "enterocoel theory", which originated in the last century, the attempt was made to trace the secondary body cavities of the Bilateria back to the gastric pouches of the Cnidaria (Coelenterata). This hypothesis was fueled by the identification of myoepithelia in the development of secondary body cavities, but it completely lacks a structural basis. We are faced once again with the dangerous suggestive power of semantic congruences. What are called epithelial muscle cells here only share the common use of a term. The Cnidaria have a basal, narrow layer of muscle fibers in cells, which function primarily as enzyme producing or digestive cells. The cells in the lining of the coelom cavities of the Bilateria are, in contrast, primarily muscle cells, and are thus usually completely filled with myonemes.

Circulatory System

Let us now turn to the tubular transport channels of the Bilateria, which can be organized as primary and secondary body cavities.

Transport systems for the circulation of body fluids are absent in almost all taxa that have a compact organization, and also in the Nemathelminthes with a primary body cavity.

With the exception of the Nemertini (see below), the blood circulatory systems are linked to the existence of hydroskeleton-like secondary body cavities. In all invertebrate Bilateria with a coelom, including the Acrania (*Branchiostoma*), they originate as gaps in the ECM between the myoepithelia of neighboring coelom cavities; they themselves have no cellular lining and therefore represent the state of primary body cavities. An epithelial lining of the blood vessels developed first in the evolution of the vertebrates.

This brings us to the Nemertini, which are an exception in two respects. The Nemertini are the only Bilateria with a compact organization which have a circulatory system. They are also the only invertebrate Bilateria whose vessels are lined with a nonmuscular epithelium (Fig. 96D). In the ground pattern of the Nemertini, there are two lateral vessels made of endothelial cells which are embedded in the ECM and the musculature (p. 205). True to definition, this condition is a secondary body cavity. In the evolutionary assessment, however, the circulatory system of the Nemertini is an autapomorphy which evolved independently of homonymous systems in the stem lineage of the Nemertini.

The lack of blood vessels in the Gnathostomulida, Plathelminthes, and Nemathelminthes is a plesiomorphous state. A corresponding transport system does not form a part of the ground pattern of the Bilateria.

Body Wall Musculature

A body wall musculature consisting of subepidermal circular and longitudinal muscles first appeared in the stem lineage of the Bilateria during the phylogenesis of the Metazoa. Its evolution would seem to be related to the origin of bilateral symmetry and unidirectional locomotion. As an autapomorphy of the Bilateria, the simplest state – one layer of loosely arranged outer circular muscle cells and a second layer of inner longitudinal muscles – belongs to the ground pattern.

Nervous System – Sensory Organs

A rostral sensory pole and a brain-like center of the nervous system evolved in order to control the activities of a bilaterally symmetrical organism. Monociliated sensory cells were adopted by the Bilateria from the ground pattern of the Eumetazoa and they became concentrated at the anterior end. The brain of the stem species of the Bilateria could not have been more than a basiepithelial accumulation of nerve cells, similar to that of the brain ganglion (frontal ganglion) of the Gnathostomulida. The first condensations leading to longitudinal nerves may have emerged in the plesiomorphous, peripheral nerve plexus.

Nephridial Organs

Organs with osmoregulatory and excretory functions occur in the Bilateria in two different forms (Figs. 47,48) – as protonephridia, and as metanephridial systems with metanephridia and podocytes (RUPPERT and SMITH 1988; BARTOLOMAEUS and AX 1992; BARTOLOMAEUS 1993a).

1. Protonephridia

   Protonephridia are proximally closed canals between the ectoderm and endoderm. They lead outward through the epidermis.

   Protonephridia appear in the Bilateria in combination with all three forms of the mesoderm/ body cavities:
   a) in unities with a compact organization such as in the Gnathostomulida, Plathelminthes, Nemertini, or Gastrotricha;
   b) in connection with a primary body cavity, such as in the Rotatoria and Acanthocephala, and in the Kinorhyncha or Priapulida;
   c) even in unities with a secondary body cavity as adults (Annelida) or in their acoelomate larvae (Annelida, Mollusca, Phoronida).

   Due to their congruences, which include details in the ultrastructure, the protonephridia of all Bilateria (in both adult organisms and larvae) can be hypothesized as homologous organs.

In the primary state, protonephridia are paired organs of ectodermal origin made up of three monociliated cells. A closed terminal cell, a canal cell, and a nephropore cell in the epidermis surround a tubular lumen (Figs. 47A,49):

a) In the terminal cell, the cilium is surrounded by eight long microvilli. They stabilize the cavity for the free movement of the cilium. The terminal cell forms the filter apparatus of the protonephridium. It is distally elongated to a hollow cylinder, which is pierced with slits or pores. The ECM is positioned on the outside of this porous structure as an ultrafilter.

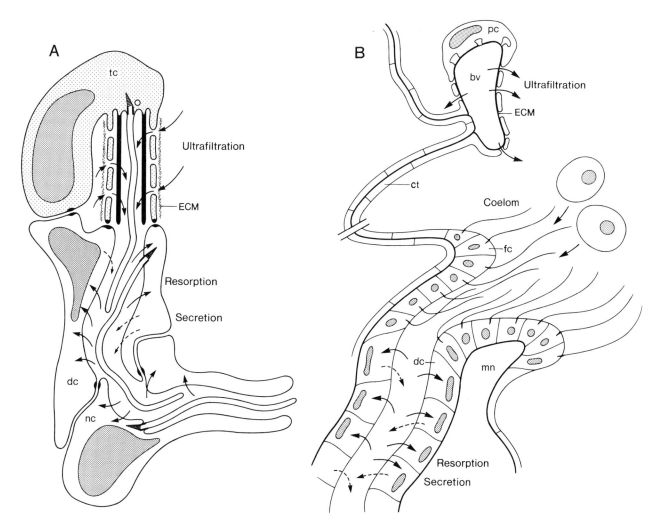

Fig. 47. Functional morphology of the protonephridium and the metanephridial system. A. Protonephridium composed of three monociliary cells in the ground pattern of the Bilateria. Ultrafiltration results when fluid passes the extracellular matrix (ECM), which lies on the perforated, hollow cylinder of the terminal cell. The fluid which has entered the cavity is pushed into the canal cell by the cilium of the terminal cell; a ring of eight microvilli (black rods) creates a constant lumen for its movement. Cilia of the canal cell and the nephropore cell continue the flow of water. These cells modify the ultrafilter through secretion and resorption. B. Metanephridial system composed of metanephridia and podocytes. The funnel of the metanephridium originates either from canal cells of the protonephridium (Annelida) or from the coelomepithelium (Phoronida). Ultrafiltration results when fluid passes the extracellular matrix (ECM) from blood vessels, on which there are processes of podocytes, and enters into the secondary body cavity. The ultrafiltrate is led outwards from here through the metanephridium, causing its modification. Metanephridia also function as gonoducts (indicated by two egg cells before the ciliated funnel). bv = blood vessel. ct = coelothel. dc = canal cell. fc = funnel cell. mn = metanephridium. nc = nephropore cell. pc = podocyte. tc = terminal cell. (Bartolomaeus and Ax 1992)

The molecular sieve of the extracellular matrix is the barrier between the intercellular body fluid and the lumen of the protonephridium.

Terminal cells with this type of construction are found in the Gastrotricha, the Gnathostomulida, and in the Actinotrocha larvae of the Phoronida.

b) A canal cell with one cilium, and

c) a corresponding nephropore cell in the epidermis characterize the protonephridia of the Gastrotricha (Fig. 49A).

The nephridial organ of the Gnathostomulida deviates only slightly from this pattern. Cilia are absent in the canal cells and nephropore cells (Fig. 49B).

An accessory centriole is present primarily next to the ciliary rootlet in the monociliated cells of the protonephridia. A diplosome is found in the aciliated cells of the Gnathostomulida.

Terminal cells, canal cells and nephropore cells are connected with each other by zonulae adherentes.

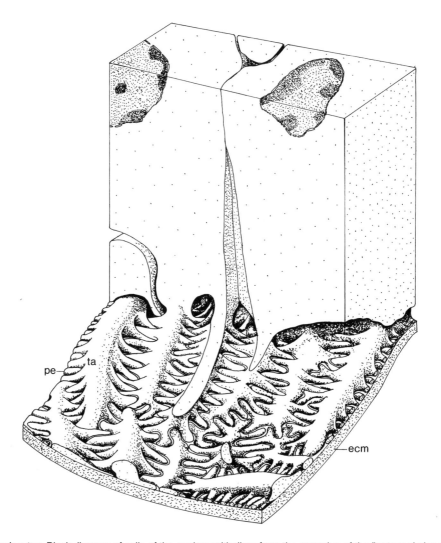

Fig. 48. Podocytes. Block diagram of cells of the coelomepithelium from the sacculus of the "antennal gland" of the American lobster *Orconectes limosus*. Long off-shoots and pedicels of the podocytes interlock on the ultrafilter of ECM which surrounds blood vessels. ecm = extracellular matrix. pe = pedicel. ta = off-shoot. (Kümmel 1977)

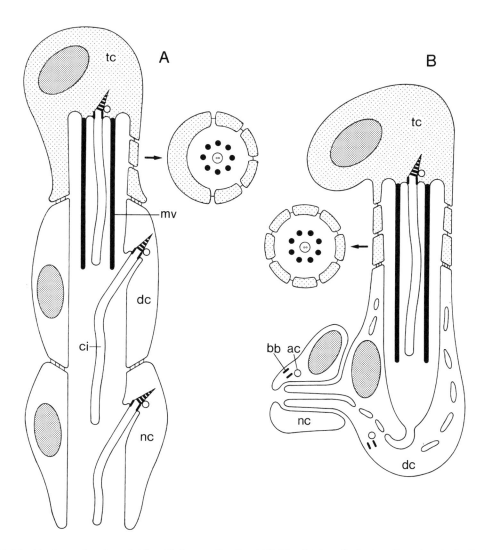

Fig. 49. Original features of protonephridia with the construction of three cells. Longitudinal sections; terminal cell with ring of eight microvilli additionally shown in a cross section. A. The protonephridium of the Gastrotricha represents the state in the ground pattern of the Bilateria. Terminal cell, canal cell, and nephropore cell each have one cilium with a ciliary rootlet and accessory centriole. B. In the Gnathostomulida the cilia are reduced in the canal and nephropore cell; basal body and accessory centriole are kept. ac = accessory centriole. bb = basal body. ci = cilium. dc = canal cell. mv = microvilli. nc = nephropore cell. tc = terminal cell. (Bartolomaeus and Ax 1992)

Variations in the basic form outlined above include the development of multiciliated cells, the reduction of microvilli or the multiplication of terminal cells, canal cells and nephropore cells. Since they are assessable as apomorphies, they can be quoted on case by case as basis for the justification of monophyletic subtaxa of the Bilateria.

2. Metanephridial Systems
Metanephridial systems are composed of metanephridia and podocytes. They are found only in correlation with secondary body cavities (Mollusca, Annelida, Sipunculida, Echiurida, Pogonophora, Vestimentifera, Phoronida, Brachiopoda, Deuterostomia).

Metanephridia are tubules located between the ectoderm and endoderm which begin proximally with a ciliated funnel that opens in a secondary body cavity; they lead outward through the epidermis. (Fig. 47B).

Podocytes are single cells positioned on the extracellular matrix surrounding the blood vessels. Basally they have long off-shoots, which branch further into delicate processes. The pedicels of neighboring podocytes closely interlock with each other, leaving only tiny slits between them (Fig. 48).

In metanephridial systems, the off-shoots and pedicels of the podocytes carry the ultrafilter made of ECM. High pressure in the blood vessels and low pressure in the body cavity cause the blood plasma to be filtered through the ECM. The coelom functions as an intermediate storehouse for the ultrafiltrate, which is modified in the last step and then excreted by the metanephridia.

In sharp contrast to the protonephridia, the metanephridia cannot be interpreted as homologous organs throughout the Bilateria. This statement results from an analysis of the relationships between the nephridial organs, where they occur together or successively.

Protonephridia – Metanephridia

The study of the ontogenetic connections between the protonephridia of larvae and the metanephridia of postlarval stages has revealed fundamental differences in the development and organization of the metanephridia in the Phoronida and the Annelida.

P h o r o n i d a . The Actinotrocha larva of *Phoronis muelleri* has paired protonephridia. Every organ consists of three complexes, each containing 30 closed, monociliated terminal cells, a canal made up of numerous monociliated cells, and the nephropore in the epidermis. During metamorphosis, the terminal cells and a proximal section of the canal are cast off and degenerate. The remaining protonephridial canal then connects to monociliated myoepithelial cells of the coelom wall which join together to form a ciliated funnel (Fig. 50).

A n n e l i d a . Trochophora larvae have paired protonephridia which, as transitory organs, are modified or disintegrated during the course of ontogenesis. Postlarval stages of development have segmentally arranged nephridial organs – either protonephridia or metanephridia.

In species of different taxa (Phyllodocidae, Alciopidae, Glyceridae, Nephtyidae), protonephridia remain the only organs of excretion for the entire life cycle. Structural similarities between the protonephridia of larvae and adults must be the expression of one and the same genetic information.

In the case of the Annelida, metanephridia stem from the metamorphosis of protonephridia. Stages of this evolution can be seen within the taxon. *Pisione remota* (Pisionidae) and *Anaitides mucosa* (Phyllodocidae) belong to annelid taxa with segmental protonephridia. At the time of sexual maturity, a ciliated funnel for the release of gametes develops from proliferating canal cells next to closed terminal cells (Fig. 51). After reproduction, the funnel is reduced. A further evolutionary step can be seen in *Pholoe minuta* (Sigalionidae). As an adult, this species has segmental metanephridia. During its development, however, it passes through an intermediary protonephridial stage with monociliated terminal cells. The ciliated funnel only develops after this stage from the separation of the proximal canal cells while the protonephridial terminal cells degenerate (Fig. 52).

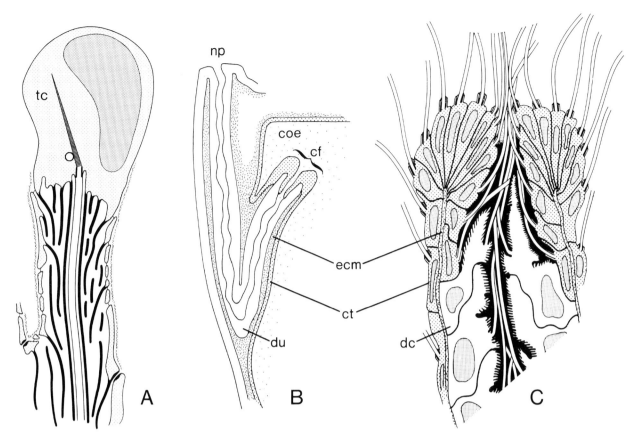

Fig. 50. Nephridial organs of *Phoronis muelleri* (Phoronida). A. Monociliary terminal cell of the protonephridium of the Actinotrocha larva shown in a longitudinal section. B-C. Metanephridium of the adult in the composition of an ectodermal
canal and mesodermal ciliated funnel. B. Longitudinal section of the organ. C. Ciliated funnel and beginning of the canal.
During ontogenesis, the canal of the metanephridium originates from the protonephridial canal of the Actinotrocha larva.
On the other hand, the terminal cells of the larval protonephridium are eliminated; a metanephridial funnel composed of
coelom cells (dotted) appears in its place in the adult. cf = ciliated funnel. coe = coelom. ct = coelothel. dc = canal cell. du
= canal. ecm = extracellular matrix. np = nephropore. tc = terminal cell. (Bartolomaeus and Ax 1992)

The metanephridia of the Annelida always develop from uniform nephridial "anlagen", and, unlike
the processes occurring in the Phoronida, the coelom epithelium plays no part in this development.

Conclusion

Paired protonephridia of ectodermal origin are a part of the ground pattern of the Bilateria. They
evolved in a monophasic, compact organism without a body cavity. The primary construction of
the protonephridium includes the following features: three-part organization made up of a terminal
cell, a canal cell, and a nephropore cell in the epidermis. Each cell has a cilium with one rootlet
and an accessory centriole. The cilium is surrounded by eight microvilli in the terminal cell; a
distal cylinder carries the ultrafilter made of ECM. Protonephridia are homologous organs in all
Bilateria, irrespective of whether they are realized in larvae or adults.

Metanephridia developed within the Bilateria from protonephridia. This evolution occurred exclusively in conjunction with the evolution of secondary body cavities. Fundamental differences in

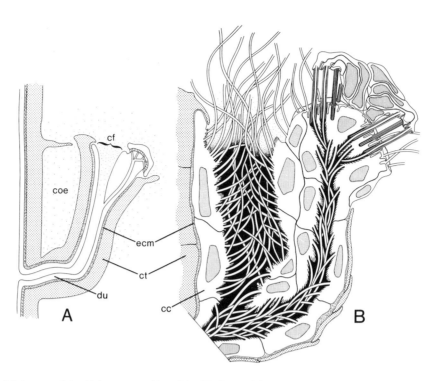

Fig. 51. Nephridial organ of *Anaitides mucosa* (Annelida, Phyllodocida). A. Longitudinal section of the organ. B. Proximal part with closed terminal cells (right) and an additional ectodermal funnel. The adult has segmental protonephridia with monociliary terminal cells and multiciliary canal cells its whole life. During sexual maturity, the canal breaks open next to the terminal cell complex and forms a multiciliary funnel composed of canal cells for the release of gametes. cc = canal cell. cf = ciliated funnel. coe = coelom. ct = coelothel. du = canal. ecm = extracellular matrix. (Bartolomaeus and Ax 1992)

ontogenesis lead to the assumption that the metanephridia of Annelida and Phoronida are not homologous. Metanephridia and the metanephridial systems evolved in the Bilateria at least twice independently, once within the Annelida, and once within the stem lineage of the Radialia. Arguments offered in support of the hypothesis that the metanephridia of certain taxa are not homologous also support the hypothesis of a multiple, convergent evolution of the respective secondary body cavities (p. 117).

Gametes – Development

The fact that numerous subtaxa of the Bilateria are composed of gonochorists, and that a free discharge of gametes into water is very common, speak in favour of including these two features as plesiomorphies in the ground pattern of the Bilateria. This applies also to the radial cleavage adopted from the Cnidaria, though this entails a problem that we shall examine a little later in more detail.

Lastly, there is no indication that diverse taxa of the Bilateria (Gnathostomulida, Plathelminthes or Gastrotricha, Nematoda, Kinorhyncha, Loricifera) were ever anything other than organisms of the benthal with a monophasic life cycle. This also holds true for the primary life-style of the Nemertini; the pilidium larva originated first within this unity (p. 210). In other words, there are valid reasons to postulate an organism for the stem species of the Bilateria that developed directly and did not include a plankton larval stage.

124

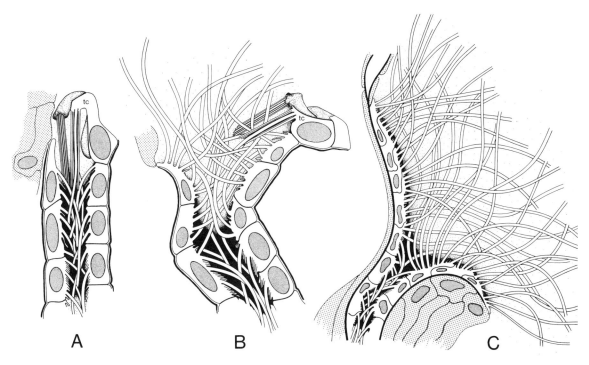

Fig. 52. Development of the metanephridium of *Pholoe minuta* (Annelida, Sigalionida) through an intermediary protone-phridium. A. Beginning of the nephridial organ as a protonephridium with monociliary terminal cells (tc) in the second segment before the pygidium. B. Opening of the proximal canal cells while the corresponding body section becomes the fourth prepygidial segment. C. Formation of the ciliated funnel composed of multiciliary canal cells and complete reduction of the protonephridial terminal cells while reaching segment position six before the pygidium. (Bartolomaeus and Ax 1992)

Summary of Features of the Ground Pattern

The stem species of recent Bilateria has had the following features (P = plesiomorphy, A = autapomorphy):

– Millimeter-sized organism (P).

– Bilaterally symmetrical body with conical ends (A).

– Inhabitants of the benthic zone (P).

– Ectoderm and endoderm with monociliated cells (P).

– Intestine without anus, and with a ventral oral opening in the anterior part of the body (P).

– Condensation of the ECM to a basal lamina beneath the ectoderm and endoderm (A).

– Compact organization without a body cavity (P).

– Body wall musculature consisting of outer circular muscles and inner longitudinal muscles (A).

– Central nervous system with brain at the anterior end (A).

- Monociliated sensory cells (P).

- Paired protonephridia of ectodermal origin. They are made up of three monociliated cells: terminal cell, canal cell and nephropore cell. Terminal cell with eight microvilli around the cilium and a porous cylinder as the carrier of the ultrafilter of ECM (A).

- Gonochoric organism (P).

- Free discharge of gametes and external fertilization in water (P).

- Radial cleavage (P).

- Direct development. Monophasic life cycle without larvae (P).

Systematization

The classical division of the Bilateria into the Protostomia and the Deuterostomia (GROBBEN 1908) is based on the fate of the blastopore in ontogenesis. Those unities in which the definitive mouth develops from the blastopore form the Protostomia. Bilateria, however, in which the blastopore region becomes the anus, and the oral opening of the adult is a new formation, are classified as Deuterostomia.

The alternative "Protostomia" – Deuterostomia is one of those dichotomous, large divisions which we reject in the phylogenetic system (p. 47). Here, too, a paraphylum is grouped next to a monophylum, just as is the case of the alternative "Protozoa" – Metazoa. The mouth of the "Protostomia" corresponds to the one primary body opening in the ground pattern of the Eumetazoa, as is realized in the Cnidaria, the Ctenophora, and in the Plathelminthomorpha. Once again, a plesiomorphy was used in this case for the purpose of classification. Since this does not justify a kinship, a taxon "Protostomia" should be eliminated from the system of the Bilateria. In the other part of the alternative, the apomorphous state of a differentiation of the blastopore region to form the anus can certainly be used to establish the monophylum Deuterostomia (p. 130).

For a consistent phylogenetic systematization of the Bilateria, two concurring hypotheses were developed in our work group (EHLERS; BARTOLOMAEUS; AX 1989c). They have been hierarchically tabulated as follows:

Hypothesis 1	**Hypothesis 2**
Bilateria	**Bilateria**
Plathelminthomorpha	**Spiralia**
Gnathostomulida	**Plathelminthomorpha**
Plathelminthes	**Euspiralia**
Eubilateria	**Radialia**
	"Tentaculata"
	Deuterostomia

The essential conflict between these hypotheses results from the problem of consistently interpreting the evolutionary origin of two clearly apomorphous features. These are the evolution of an intestine with anus, and the origin of the ontogenesis pattern of the spiral quartet 4d cleavage.

The Plathelminthomorpha, with the sister groups Gnathostomulida and Plathelminthes, form the only large unity of the Bilateria with a blind intestine without an anus. In comparison to this plesiomorphy (p. 131), the presence of an anus in all other Bilateria is the usual, apomorphous, case. Secondly, the Plathelminthomorpha undergo spiral cleavage, which can clearly be proved in comparison with the radial cleavage of the Cnidaria to be an apomorphy.

If we follow the principle of the most parsimonious explanation, then the next step is to attempt to interpret both the apomorphies – anus and spiral cleavage – as products of a single evolutionary process.

Hypothesis 1 corresponds to this methodological postulate. All the other Bilateria are positioned opposite the Plathelminthomorpha as the sister group Eubilateria – assuming that a continuous intestinal tubule with anus originated once in the stem lineage of a monophylum Eubilateria. This assumption makes it necessary to state exactly when the spiral quartet 4d cleavage evolved. According to hypothesis 1, it must have originated in the stem lineage of the Bilateria, and must hence form a feature in the ground pattern of the Bilateria. Or more precisely, all deviating cleavage patterns within the Bilateria must be interpretable as the results of modifications of the spiral quartet 4d cleavage. This is indeed true for the discoidal cleavage of the eggs of squid which are rich in yolk, because the Cephalopoda are part of the monophylum Mollusca which have the spiral quartet 4d cleavage in their ground pattern. It also holds true for the superficial cleavage pattern within the Arthropoda, because these, together with the Annelida and other unities, form a monophylum Articulata, and a corresponding spiral cleavage can be proven to be part of its ground pattern. It is, however, definitely not the case in the radial cleavage of the "Tentaculata" and the Deuterostomia. We do not have the slightest justification to consider the radial cleavage of these Bilateria groups as a secondary change of the spiral quartet 4d cleavage pattern (DOHLE 1989).

This consideration leads us to **hypothesis 2**, which systematizes a unity Spiralia (Plathelminthomorpha, Nemertini, Kamptozoa, Mollusca, Articulata) and a unity Radialia ("Tentaculata", Deuterostomia) as adelphotaxa of the Bilateria. This hypothesis postulates the adoption of radial cleavage from the Cnidaria into the ground pattern of the Bilateria and the continuation of this plesiomorphous ontogenesis pattern in the stem lineage of the Radialia. On the other hand, the spiral quartet 4d cleavage evolved from the original radial cleavage once in the stem lineage of the sister group Spiralia (Figs.53, 54).

This second hypothesis would mean that the apomorphous feature anus must have originated twice convergently – once in the stem lineage of the Radialia, and once within the Spiralia in the stem lineage of a unity Euspiralia, which is positioned, as the sister group, next to the Plathelminthomorpha.

When considering the alternatives listed, the postulate of an intestine with anus that evolved twice convergently appears to be the simpler solution compared with an evolution of the radial cleavage in the Radialia from the pattern of the spiral quartet 4d cleavage. For this reason, we base the systematization of the Bilateria (Fig. 55) on hypothesis 2 with the adelphotaxa Spiralia – Radialia.

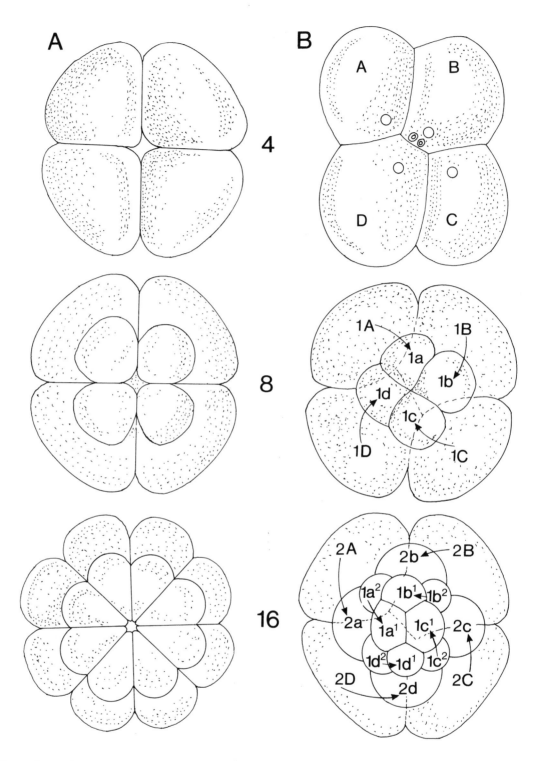

Fig. 53. Comparison of the original radial cleavage of the Metazoa with the derived spiral quartet cleavage of the Spiralia. View of the animal pole. A. Radial cleavage. In the stage of four cells, the blastomeres lie on one level next to each other. The change to the eight-cell stage results through equatorial divisions where the spindle axis of the division figures are parallel to the polar axis of the germ; the consequences are two circles of four blastomeres with an exact covering of cells and cleavage in the verticle. Stage 16 is reached through meridional division of the blastomeres; the result is a radial symmetrical germ in which two circles each containing eight blastomeres lie directly on top of each other. B. Spiral quartet cleavage. In stage 4 the new, usually smaller blastomeres, whose spindle axes lean to the right towards the polar axis of ▶

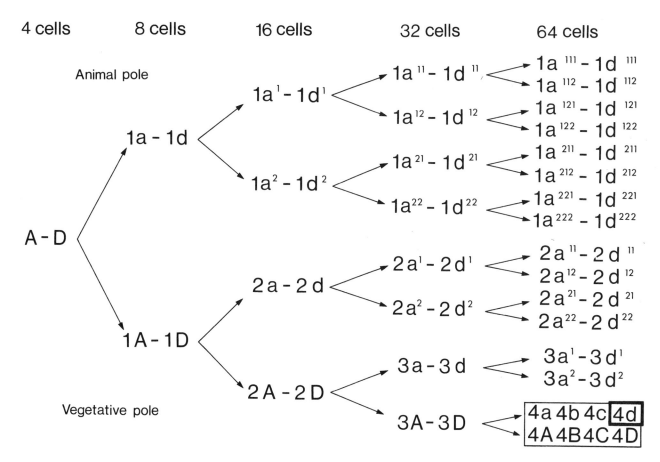

4 cells	8 cells	16 cells	32 cells	64 cells

Animal pole

Vegetative pole

$1a^{11}-1d^{11}$
$1a^1-1d^1$
$1a^{111}-1d^{111}$
$1a^{112}-1d^{112}$

$1a-1d$
$1a^{12}-1d^{12}$
$1a^{121}-1d^{121}$
$1a^{122}-1d^{122}$

$1a^{21}-1d^{21}$
$1a^{211}-1d^{211}$
$1a^{212}-1d^{212}$

$1a^2-1d^2$
$1a^{22}-1d^{22}$
$1a^{221}-1d^{221}$
$1a^{222}-1d^{222}$

$A-D$

$2a^1-2d^1$
$2a^{11}-2d^{11}$
$2a^{12}-2d^{12}$

$2a-2d$
$2a^2-2d^2$
$2a^{21}-2d^{21}$
$2a^{22}-2d^{22}$

$1A-1D$

$3a-3d$
$3a^1-3d^1$
$3a^2-3d^2$

$2A-2D$

$3A-3D$
4a 4b 4c 4d
4A 4B 4C 4D

Fig. 54. Markings of the individual blastomeres of spiral quartet cleavage up to the 64th cell stage. In the Annelida, these 64 cells have the following prospective meaning: 56 cells ($1a^{111}-1d^{222}$, $2a^{11}-2d^{22}$, $3a^1-3d^2$) produce ectoderm and ectodermal derivates such as the nervous system and protonephridia. The endoderm originates from the seven cells, 4a,b,c, and 4A-D. The blastomere 4d is the mesoderm cell.

the germ, are severed; the four micromeres 1a-d in stage 8 arrange themselves above the cleavages between the macromeres 1A-D. During the change to stage 16 the spindle axis lean to the left; since both circles release new micromeres upwards at an angle, the 16 blastomeres arrange themselves in four layers, one upon another.

The radial cleavage is characterized by the alternation between equatorial and meridional division of the blastomeres; a layer of blastomeres is released to the animal pole with only every second step of cleavage. During spiral cleavage, however, new blastomeres are pushed to the top with every new step in the change of dexiotrop and leiotrop divisions. (Siewing 1969)

Spiralia – Radialia

Let us first take a brief look at the precise meaning of a juxtapositioning of Spiralia and Radialia. There is an apomorphous congruence between the Plathelminthomorpha[6], Nemertini[6], Kamptozoa, Mollusca and Articulata in the spiral cleavage which has the following characteristics: (1) alternating dexiotropous and leiotropous formation of four micromere quartets as a result of a corresponding inclination of the spindles; (2) origin of the mesoderm from the micromere 4d.

A repeated convergent evolution of this complex cleavage pattern is highly unlikely. The spiral quartet 4d cleavage is valid as a convincing autapomorphy for the justification of a monophylum Spiralia.

On the other hand, the plesiomorphous radial cleavage cannot be used to justify the sister group Radialia. The group name introduced by JEFFERIES (1986) is, for this reason, poor, but we will have to accept it as an existing name.

The monophyly of a unity from the "Tentaculata" taxa Phoronida, Brachiopoda, Bryozoa, as well as the Deuterostomia, can be justified on the basis of their life-style as tube dwellers and correlated apomorphous characteristics. The transition from a vagile surface dweller to a sedentary organism – probably to a tube-forming species comparable with the Phoronida or the Pterobranchia – occurred in the stem lineage of the Radialia. Tentacles evolved in conjunction with this transition at the anterior end, and a secondary body cavity evolved with a few compartments. It remains uncertain as to whether two or three coelom compartments are part of the ground pattern of the Radialia (BARTOLOMAEUS 1993a).

To summarize: a sedentary life-style, tentacles and a hydroskeleton which evolved convergently to the coelom of the Annelida (p. 116) form the autapomorphies of the Radialia in their interpretation as adelphotaxon of the Spiralia (Fig. 55, autapomorphy block 4). We must add the unavoidable assumption of the evolution of an anus in the stem lineage of the Radialia, convergent to the origin of proctodeum and anus within the Spiralia (p. 132). The "Tentaculata" are not justifiable as a monophylum within the Radialia. The origin of the mouth as a new formation and development of the anus from the blastopore (Fig. 55, autapomorphy 5) characterize the Deuterostomia.

The very large Bilateria taxon of the **Nemathelminthes** (Aschelminthes) has not yet been satisfactorily systematized. Even the monophyly for a unity of the taxa Gastrotricha, Nematoda, Nematomorpha, Priapulida, Kinorhyncha, Loricifera, Rotatoria and Acanthocephala is hard to justify. The brain in its state as a circumintestinal ring, and the granular basal layered cuticle with a monolamellar, membrane-like epicuticle have been proposed as autapomorphies for all but the latter two of these taxa (NEUHAUS 1994). The terminal position of the mouth at the tip of the body could be another derived feature of the Nemathelminthes.

We will characterize the Nemathelminthes and their subtaxa after the following analysis of the Spiralia, and will then argue about the position of the Nemathelminthes within the Bilateria.

---

6 More precise studies are desirable.

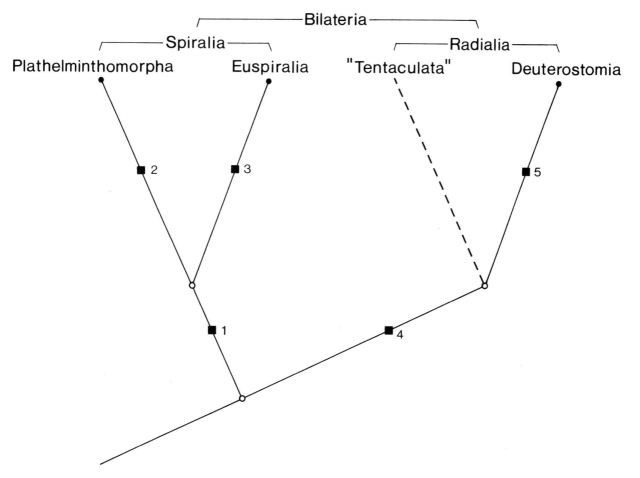

Fig. 55. Diagram of phylogenetic kinship relations within the Bilateria based on the hypothesis of an adelphotaxa relation between the unities Spiralia and Radialia.

Spiralia

Plathelminthomorpha – Euspiralia

Our next step is to justify the hypothesized sister group relation between the Plathelminthomorpha and all other Spiralia, which have been united here as Euspiralia (Fig.56).

It is important in this context to remember that the Plathelminthomorpha (Gnathostomulida + Plathelminthes) are Bilateria without an anus, and that we have assessed this state to be a plesiomorphy (p. 112, 127).

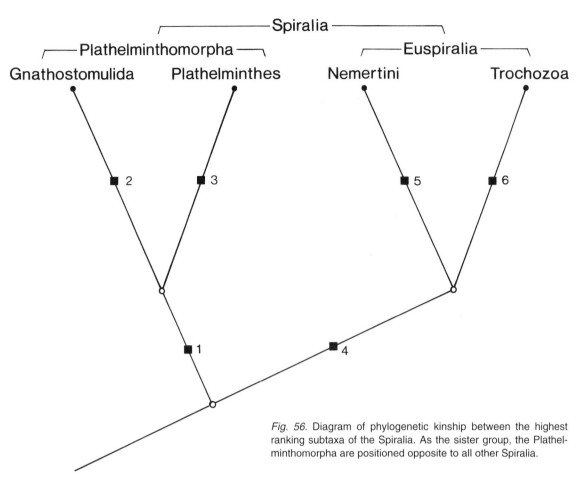

Fig. 56. Diagram of phylogenetic kinship between the highest ranking subtaxa of the Spiralia. As the sister group, the Plathelminthomorpha are positioned opposite to all other Spiralia.

Applying the principle of parsimony, we then go on to postulate a single evolution within the Spiralia of the continuous intestine with anus, or, more precisely, with a dorsal anal pore (BARTOLOMAEUS 1993b), in the stem lineage of the Euspiralia. We identify this as an autapomorphy which justifies the Euspiralia as a monophylum (p. 199).

On the other hand, hermaphroditism, internal fertilization, and perhaps also the inability of differentiated somatic cells to divide must be assessed to be autapomorphies of the Plathelminthomorpha.

Plathelminthomorpha

■ **Autapomorphies**
(Fig. 55 → 2; 56 → 1)

– Hermaphroditism.

– Direct transfer of sperm and internal fertilization of the egg cells.

– Thread-like sperm.

– Lack of mitosis in somatic cells.

Gonochorism, postulated as part of the ground pattern of the Bilateria (p. 124), is seen in the majority of the Spiralia taxa. Hence, the hermaphroditism which is consistently realized in the Gnathostomulida and the Plathelminthes must be interpreted as an apomorphy. A change from gonochorism to hermaphroditism that occurred once in the stem lineage of the Plathelminthomorpha can be hypothesized without conflict.

The free discharge of sperm and external fertilization are prevalent throughout the Spiralia, including the Annelida. The derived phenomena of a direct transfer of sperm and the fertilization of egg cells within the mate are correlated with a further apomorphy of the Plathelminthomorpha found in the thread-like, elongated sperm without a distinct acrosomal complex. This can be assessed with certainty to be a derived feature, contrasting with the primary sperm pattern of external fertilization within the Metazoa (p. 54). Thread-like sperm with a terminal cilium are found in only a few taxa of the Plathelminthomorpha – in the Filospermoidea (*Pterognathia, Haplognathia*) within the Gnathostomulida, and in the Nemertodermatida (*Nemertoderma, Meara*) within the Plathelminthes. In both unities the sperm undergoes various changes.

One characteristic of the Plathelminthes worthy of mention is the inability of somatic cells to divide (EHLERS 1985; BAGUÑA et al. 1989; PALMBERG 1990). Worn-out cells are replaced by undifferentiated stem cells found in the gaps between the ectoderm and endoderm. These neoblasts are the basis of the high regenerative ability of the Tricladida (planarians). The lack of mitosis in differentiated body cells is part of the ground pattern of the Plathelminthes.

What about this phenomenon in the sister group Gnathostomulida? Mitotic divisions of epidermal cells or other somatic cells have not yet been observed, not even in young individuals. Small, ovoid cells in basiepithelial position in *Haplognathia rosea* may be stem cells (replacement cells) which give rise to epidermal cells with a locomotory cilium (LAMMERT 1989, 1991). In the plathelminth *Rhynchoscolex simplex* (Catenulida), too, undifferentiated stem cells between the basal regions of multiciliated epidermal cells have been observed (EHLERS 1992b).

This phenomenon has not been adequately analyzed in the Gnathostomulida, however. The lack of mitosis in differentiated somatic cells can be considered only with reservation to be part of the ground pattern of the Plathelminthomorpha. The characteristic feature might even have first evolved in the stem lineage of the Plathelminthes.

Gnathostomulida – Plathelminthes

The name Gnathostomulida was based on the feature of a pair of cuticular jaws in the pharynx. The jaws and an unpaired basal plate form the essential autapomorphy of the taxon.

More exciting, however, is the construction of the ectoderm from monociliated cells, which is a very old plesiomorphy adopted by the Spiralia from the ground pattern of the Metazoa. Each of the epidermal cells over the entire body has only one single cilium with an accessory centriole next to the basal body. No other unity of the Bilateria has this kind of skin with a locomotory function, including the sister group of the Plathelminthes.

This brings us to the autapomorphies of the flatworms. Multiciliated epidermal cells evolved in the stem lineage of the Plathelminthes with the reduction of the accessory centriole of the cilia. This process of multiplication also occurred in the Protonephridia; the terminal cells have at least two cilia without an accessory centriole.

If our hypothesis of an adelphotaxa relation Gnathostomulida – Plathelminthes is sound, then the multiciliation of the epidermis in the Euspiralia must have originated convergently with their evolution in the Plathelminthes.

Gnathostomulida

■ **Autapomorphies**
(Fig. 56 → 2)

- Two cuticular jaws and a basal plate in the pharynx.

- Spiral cilium receptors in the anterior end.

- Cross-striated musculature.

- Protonephridia.
  Several organs serially arranged. Canal cells and nephropore cells without cilia.

- Intestinal cells without cilia.

Based on the description of the first two species, *Gnathostomula paradoxa* from the North and Baltic Seas and *Gnathostomaria lutheri* from the Mediterranean (Ax 1956b), the number of gnathostomulid species has now increased to about 80 (Sterrer 1991). The millimeter-sized organisms lives worldwide in the interstices of sandy ocean floors.

Sterrer (1972) divided the Gnathostomulida into the Filospermoidea (*Haplognathia, Pterognathia*), and the Bursovaginoidea (*Gnathostomaria, Gnathostomula, Austrognathia* et al.). Lammert (1986) argued that each of these two unities could be a monophylum, and that they, as sister groups, form the highest-ranking subtaxa of the Gnathostomulida. Let us take a brief look at the facts in favor of this argument. This will take us some way into the feature pattern of the Gnathostomulida, which we will look at in more detail later.

In the **Filospermoidea**, the *Haplognathia* and *Pterognathia* species are usually several millimeters long, extremely thin, and equipped with a very long rostrum (Fig. 57). In contrast to this, the Bursovaginoidea, around 1–2 mm, have a comparatively short anterior end in front of the ventral mouth (Fig. 58), corresponding to the condition we are assuming to be part of the ground pattern of the Plathelminthes. The out-group comparison with the adelphotaxon permits the following hypothesis. The filiform body with a long rostrum is a derived state within the Gnathostomulida, which can be interpreted as an adaptation to the interstitial environment in sand. Diverse subtaxa of the Plathelminthes underwent similar modification. In the context of the Gnathostomulida, the thread-like habitus with a rostrum is therefore an autapomorphy of the Filospermoidea.

The distinct characteristic features of the **Bursovaginoidea** are not difficult to demonstrate. Only species combined under this name have a sensorium with long tactile bristles on the anterior end. The Bursovaginoidea all have aciliated sperm, and also the presence of a bursa as a storage organ for the sperm of the mate is certainly an autapomorphy.

Ground Pattern

Monociliation is a plesiomorphous feature of the Gnathostomulida. There are no multiciliated cells in this taxon. The entire epidermis of the body is ciliated. Each locomotory cell has a long cilium with the original 9 x 2 + 2 pattern of microtubuli. The base of the cilium is in a ciliary pit beneath the cell surface from where eight short microvilli arise forming a collar around the cilium. The basal structures are composed of the basal body of the cilium, an accessory centriole, a rostral and a caudal ciliary rootlet, as well as the basal foot (Figs. 59B,C; 61A). A small appendage connects the accessory centriole with the caudal rootlet.

Monociliation also applies to **sensory organs** of ciliary origin. There is a sensorium with a few pairs of long tactile bristles at the anterior end of the Bursovaginoidea. Several monociliated receptor cells, each with a ciliary rootlet, form a single bristle.

All Gnathostomulida have characteristic spiral cilium receptors whose function is unknown (Fig. 59A). They are arranged in pairs, in the Filospermoidea in the rostrum, and in the Bursovaginoidea at the anterior end. The individual receptor cells have a cilium with a basal body and an accessory centriole, but no ciliary rootlet. The extremely long cilium is coiled up in cavities inside the receptor cell. There are no comparable sensory organs in the Plathelminthes. The spiral cilium receptor can be assessed as an autapomorphy of the Gnathostomulida.

The **nervous system**, with its predominantly intraepidermal differentiation, shows distinct plesiomorphous traits. Two ganglia appear as collections of perikarya and neurites – a brain, or frontal ganglion, in the anterior end, and a buccal ganglion in the pharynx. The Filospermoidea have a pair of longitudinal nerves. Three pairs have been found in the Bursovaginoidea.

The **musculature** is, in comparison with the Plathelminthes, of interest. While the majority of the Plathelminthes have smooth myofibrils which are part of their ground pattern, the Gnathostomulida have a cross-striated musculature. This applies to the weak body wall musculature composed of circular and longitudinal muscles, as well as to the musculature of the pharynx. The z-system is made up of isolated z-dots in the Bursovaginoidea and of thread-like z-bands in the Filospermoidea.

The organs are close to each other inside the body. There is no parenchymatous tissue.

The basal lamina is located between the epidermis and the body musculature, but not further inwards between musculature and intestine.

The **pharyngeal apparatus** is one of the distinguishing characteristic features of the Gnathostomulida (Fig. 59D).

The ventral position of the mouth in the anterior end is a plesiomorphy. It leads into an oral cavity which is formed by an epithelium made up of high, secretory cells equipped with microvilli. A tube-like esophagus arises dorsally from the oral cavity and leads into the gut lumen. A muscular pharyngeal bulb is positioned ventrally. It contains the paired jaws, the basal plate, and the buccal ganglion. The basal plate, which protrudes from the mouth as a grazing organ, and the jaws, joined to form a pincer, are cuticular structures made up of special cells. Characterization of individual species is primarily based on the differences in these structures.

The Gnathostomulida have a blind, closed **intestine**. The intestinal cells are aciliated. There are no special glandular cells.

Up to ten pairs of **protonephridia** are arranged serially from the pharynx to the posterior end. There are no connections between the individual organs. Each protonephridium is composed of a terminal cell, a canal cell and a nephropore cell. The terminal cell is the carrier of the porous filter with a layer of ECM. It has a cilium with a ciliary rootlet and an accessory centriole. Eight long microvilli surround the cilia. The canal cell with an intracellular lacunar system, and the nephropore cell have a pair of centrioles, but no cilia (Fig. 49B). The protonephridia of the Gnathostomu-

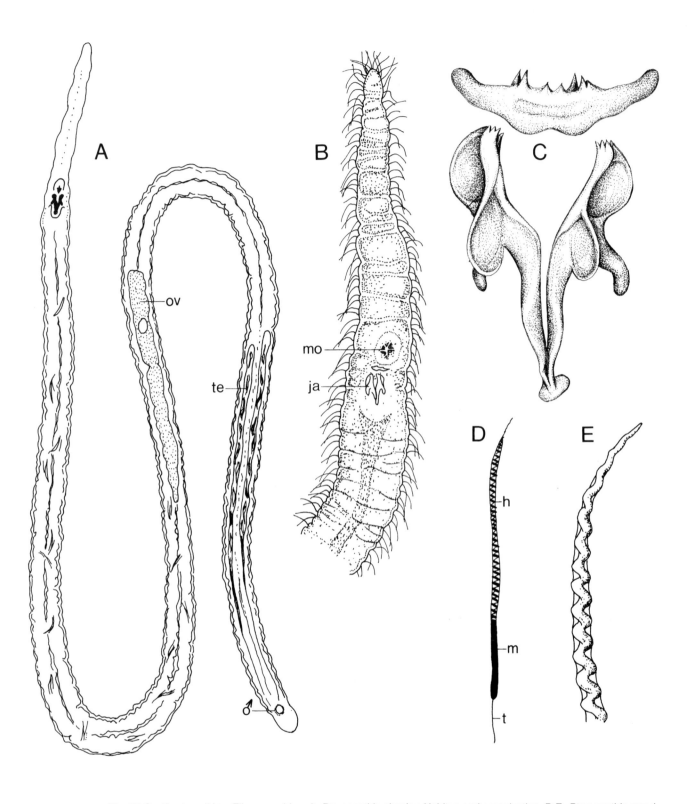

Fig. 57. Gnathostomulida – Filospermoidea. A. *Pterognathia simplex*. Habitus and organization. B-E. *Pterognathia swedmarki*. B. Anterior end with long rostrum. C. Basal plate and jaw of the pharyngeal apparatus. D. Thread-like sperm with head, middle piece, and tail. E. Head of the sperm with spiral coil. h = sperm head. ja = jaw. m = middle piece. mo = mouth. ov = ovary. t = sperm tail. te = testis. (A,B,D,E Sterrer 1968; C Müller and Ax 1971)

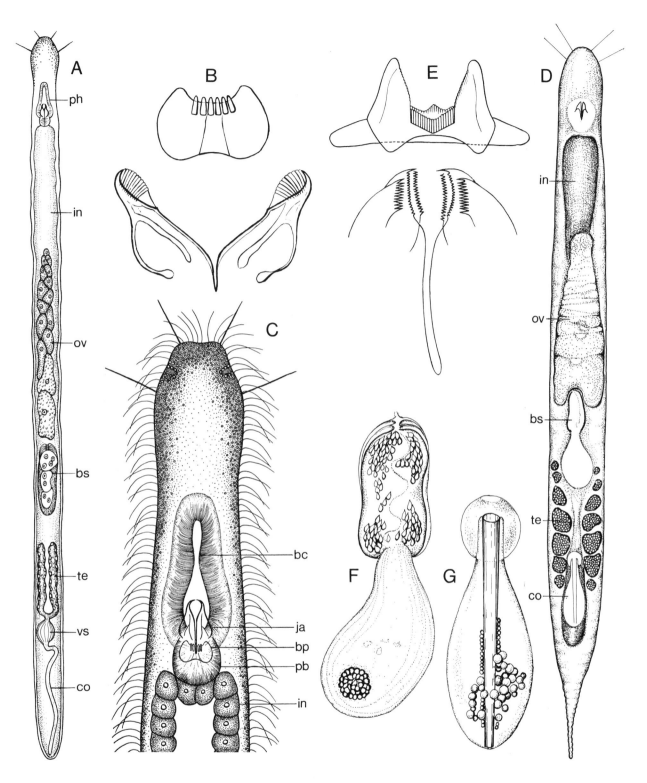

Fig. 58. Gnathostomulida – Bursovaginoidea. A-C. *Gnathostomaria lutheri.* A. Habitus and organization: B. Basal plate and jaw. C. Anterior end with pharyngeal apparatus. D-G. *Gnathostomula paradoxa.* D. Habitus and organization. E. Basal plate and jaw. F. Bursa with drop-shaped sperm from a mate. G. Copulatory organ with stylet and secretions. bc = buccal cavity. bp = basal plate. bs = bursa. co = copulatory organ. in = intestine. ja = jaw. ov = ovary. pb = pharyngeal bulb. ph = pharynx. te = testis. vs = vesicula seminalis. (Ax 1964a, 1964b, 1965; Müller and Ax 1971)

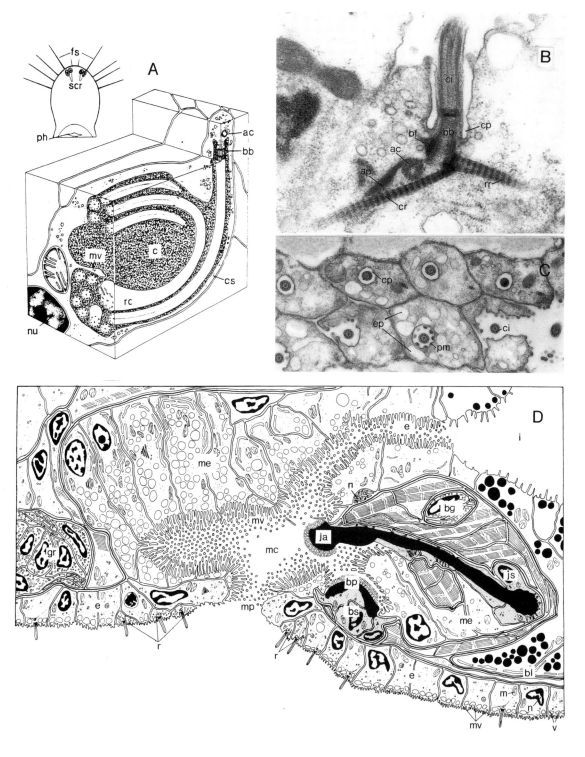

Fig. 59. Gnathostomulida. Ultrastructure. A-C. *Gnathostomula paradoxa.* A. Spiral cilium receptor. Reconstruction. Above left: Receptors positioned in the anterior end. B. Basal structures of the cilium of an epidermis cell. C. Surface section through the epidermis with one cilium per cell. D. Sagittal reconstruction of the pharyngeal apparatus of *Haplognathia simplex.* Basal plate, jaw and musculature in the pharyngeal bulb. ac = accessory centriole. ap = appendix of the accessory centriole. bb = basal body. bf = basal foot. bg = buccal ganglion. bl = basal lamina. bp = basal plate. bs = formation cell of the basal plate. c = intracellular cavity. ci = cilium. cp = ciliary pit. cr = caudal ciliary rootlet. cs = boundary of the intracellu- ▶

138

lida represent a very original state. In comparison with the structure of paired protonephridia in the ground pattern of the Bilateria, there are two apomorphies: the multiplication and serial arrangement of the organs, and the lack of cilia in the canal and nephropore cell.

The Gnathostomulida are without exception **hermaphrodites**.

The very simple structure of the male reproductive organs and the sperm in the Filospermoidea can be interpreted as original features. Paired, elongated testes are found in the second half of the body. The vasa deferentia (sperm ducts) can unite together right in front of the copulatory organ, which is merely a collection of glands surrounding the male genital pore. The Filospermoidea have thread-like sperm 100 μm long, which are pointed at the front and have a cilium (9 x 2 + 2 pattern) at the terminal end (Fig. 57D,E).

In the Bursovaginoidea, the testes can be divided into two rows of follicles (*Gnathostomula*). The most important feature, however, is the evolution of differentiated copulatory organs, for example, a soft tube-like organ in *Gnathostomaria*, or a hard stylet with eight to ten intracellular rods in a penis sheath in *Gnathostomula* (Fig. 58G). Due to their lack of cilia, the sperm of the Bursovaginoidea are highly derived, from "dwarf" sperm of 2–3 μm in length, rounded structures with foot-like protrusions (*Gnathostomula*), to the conuli (*Austrognathia*), giant mushroom-shaped sperm of up to 45 μm long.

In the case of the female reproductive organs, the ovary is very uniform throughout the taxon. An unpaired organ always lies dorsally in the middle section of the body. The entolecithal eggs grow in a caudal direction. Drops of shell substance are not produced. In addition, there are no oviducts. The mature eggs are released by the rupture of the body wall.

Whereas the Filospermoidea have no further organs apart from the ovary, a bursa for the storage of foreign sperm is a constitutive feature of the Bursovaginoidea (Fig. 58F). The bursal organ is located behind the ovary. In some cases, a vagina leads into the bursa (*Austrognatharia*).

The Gnathostomulida develop directly by means of quartet spiral cleavage. Only one *Gnathostomula* species has been studied to date (RIEDL 1969). A detailed study of their ontogenesis would be very useful.

Summary of Features of the Ground Pattern

The stem species of the Gnathostomulida has had the following features (P = plesiomorphy, A = apomorphy).

– Millimeter-sized organism with a round cross section of the body (P), possibly with a short anterior end without rostrum (P).

– Epidermis completely ciliated (P).

– Monociliated epidermal cells. Basal structures of the cilium with an accessory centriole, a rostral and a caudal ciliary rootlet (P).

– Spiral cilium receptors (A).

---

lar cavity. e = epidermis. ep = epitheliosomes. fs = frontal sensorium. i = intestine. ja = jaw. js = jaw formation cell. m = mitochondrium. mc = mouth cavity. me = secretory epithelium of the mouth cavity. mp = mouth pore. mv = microvillus. n = buccal nerve. nu = nucleus. ph = pharynx. pm = microvillus of ciliary pit. r = receptor cell. rc = coiled cilium. rr = rostral ciliary rootlet. scr = spiral cilium receptor. (Lammert 1984, 1989, 1991)

- Nervous system with intraepidermal frontal ganglion (P), with a buccal ganglion (A) and distinct longitudinal nerves (?A).

- All muscle cells with cross striated myofibrils (A).

- Compact body cavity without parenchymatous tissue (P).

- Digestive tract with ventral oral opening at the anterior end (P).

- Pharynx with basal plate and jaw (A).

- Blind intestine without anus (P) made up of aciliated cells (A).

- Protonephridia made up of a terminal cell, canal cell, and nephropore cell (P). Serial arrangement of the excretory organs (A). Lack of cilia in the canal and nephropore cell (A).

- Hermaphrodite (P). Internal fertilization (P).

- Paired testes and vasa deferentia. Simple copulation organ composed of glandular cells (P). Thread-like sperm with one cilium (P).

- Unpaired ovary (?A) with entolecithal egg formation (P). Lack of oviducts (P).

- Direct development (P). Quartet spiral cleavage (P).

Plathelminthes

■ **Autapomorphies**
(Fig. 56 → 3)

- Multiciliated epidermal cells without accessory centriole at the cilia.

- Multiciliated gastrodermal cells.

- Protonephridium: two cilia in the terminal cells and doubling of the microvilli collar from 8 to 16 (12–18). Cilia in the terminal cell again without accessory centriole.

Almost all zoology students today learn that the Plathelminthes, or flatworms, are divided into three unities of identical rank known as the "classes" Turbellaria, Trematoda, and Cestoda.
Limnetic Tricladida (planarians) are usually taken as representatives of the free-living Turbellaria in exercises. The dark, pigmented *Dugesia gonocephala*, or the milky-white *Dendrocoelum lacteum*, whose triple-branched intestine is especially suitable for demonstration purposes, are favorite study objects; and what one learns in one's youth can easily become the opinion of the respected scientist. The Tricladida continue to be considered the representatives of the flatworms when establishing the phylogenetic kinship relations of the Plathelminthes within the Metazoa

using the 18S ribosomal RNA. The American species *Dugesia tigrina* has been selected for this purpose (FIELD et al. 1988).

All this is wrong, or at least inadmissible, as the following passages will show.

(1) First of all, the division into Turbellaria, Trematoda, and Cestoda is to be rejected, because the Turbellaria are a paraphyletic collection of free-living species with a ciliated epidermis. As opposed to the parasites, we are clearly dealing in this case with plesiomorphies that are of no use for systematization purposes.
Diverse monophyletic subtaxa of free-living Plathelminthes are successively more closely related with the parasitic Trematoda + Cestoda than with other unities of free-living species, as will be shown below. In other words, there are no reasons for the existence of a taxon Turbellaria in the phylogenetic system of the Plathelminthes. It has to be eliminated along with the name Turbellaria.

(2) The Trematoda form a paraphylum in the traditional grouping of Monogenea and Digenea. Based on the common feature of sickle-shaped hooks on the posterior end of the larvae, however, the Monogenea and the Cestoda are interpretable as adelphotaxa, and are thus grouped together under the name Cercomeromorpha.

(3) Finally, in the traditional division into Turbellaria, Trematoda and Cestoda, the kinship relation of the two large parasite groups (regardless of their composition in detail) remains open. Do the Trematoda and Cestoda form a monophylum or not? This also means: did parasitism of the Trematoda and Cestoda develop once in a common stem lineage, or perhaps several times convergently? The adults in both groups have a unique skin structure, which is unmatched within the Metazoa. When the larval ciliated cells are cast off, stem cells from the inside of the body partially protrude and adhere to each other on the surface to form a new, syncytial skin. This speaks unmistakably in favor of the monophyly of the parasites, which leads to the conclusion that the Trematoda and Cestoda must be united under a proper name, the Neodermata.

(4) Can we work with the Tricladida as "typical" representatives of the Plathelminthes? The simple answer is "no", because there are several different features of the Tricladida which undoubtedly form apomorphies when compared with the feature pattern of other taxa of free-living Plathelminthes. These apomorphies include the body size of a few centimeters, a flattened body and the presence of an extensive parenchyma between the ectoderm and endoderm. The majority of free-living Plathelminthes are millimeter-sized, have a round cross section of the body, and very few parenchymatous cells. The Plathelminthes are primarily neither flat nor parenchymatous worms. The apomorphies continue with the pharynx plicatus, the intestine with three branches, the separation of the ovary into germarium and vitellarium, and the formation of ectolecithal "eggs" composed of egg and yolk cells. The embryonic pharynx, used to absorb the yolk, is yet another distinct apomorphous feature in the ontogenesis of the Tricladida.
All these features must be set aside when trying to clear up the phylogenetic relations of the Plathelminthes. Doing so, however, raises the urgent question to what extent 18S ribosomal RNA and other molecular features of the planarian *Dugesia tigrina* can have apomorphous characteristics. The situation does not become more favorable if the 18S rRNA of five further Tricladida and one species of the Rhabdocoela with germarium and vitellarium (*Bothromesostoma personatum*) are sequenced (RIUTORT et al. 1993). The issue can really be decided only by comparing the molecular patterns of a sizeable number of species from different taxa, especially of species having primary ovary and plesiomorphous endolecithal egg production. In other words, their manifestation

in the ground pattern of the Plathelminthes needs first to be reconstructed before they can be used for phylogenetic research.

The analysis of structural features has uncovered in an exemplary manner the ground pattern of the Plathelminthes right down to the ultrastructural level. A consistent phylogenetic system is available (EHLERS 1985, 1986, 1995). The taxon is ideal in a textbook on phylogenetic systematics to demonstrate the current state of research, even if space restrictions allow only mention of the especially obvious autapomorphies that characterize the various monophyla.

Ground Pattern

The feature pattern of the stem species of the Plathelminthes is only slightly more developed than the ground patterns of the Bilateria and the Plathelminthomorpha (Fig. 60). The characteristic features that are certain as autapomorphies are marked in the following outline with an (A).

The stem species of the Plathelminthes is a millimeter-sized benthic organism with a round cross section. **Multiciliated epidermal cells** (A) completely cover the body. The epidermal cilia have a rostral and a caudal root. Accessory centrioles are absent (A) (Fig. 61B, C). There is a glycocalyx between the microvilli of the epidermal cells. There are no cuticular formations. The ECM forms a basal lamina beneath the epidermis.

Monociliary sensory cells and glandular cells are located in the epidermis and also subepidermally. Processes of subepidermal sensory cells and ducts of the glands run between the epidermal cells. A smooth (nonstriated) musculature is part of the ground pattern of the Plathelminthes. This feature is common within the taxon. The subepidermal body wall musculature is composed of outer circular muscles and inner longitudinal muscles. The muscle cells are embedded in the extracellular matrix. Epithelial muscle cells do not exist in the Plathelminthes.

The stem species of the Plathelminthes has a **compact, acoelomate organization** without a body cavity. The subepidermal sensory cells and glandular cells, the muscle cells, nerve cells, protonephridia and the totipotent stem cells are embedded in ECM in the narrow gaps between the ectoderm and endoderm. Intraepidermal neoblasts may also be present in addition to the subepidermal neoblasts. There are **no parenchymatous** or **mesenchymatous cells** in the ground pattern of the Plathelminthes.

A simple oral pore connects the epidermis with the gastrodermis ventrally in the anterior half of the body. The intestine is sack-shaped without diverticula. There is no anus. Glandular cells, which produce digestive enzymes, are found in between the multiciliated intestinal cells (A), forming a single layer of cells.

A pair of protonephridia is part of the ground pattern of the Plathelminthes. Each protonephridium is composed of a minimum of three cells – a terminal cell, a canal cell, and a nephropore cell. With two cilia in the terminal cell and about 16 microvilli which form a collar around the cilia (Catenulida, Fig. 63), the number of cilia and microvilli is double (A) against that of the sister group Gnathostomulida.

The stem species of the Plathelminthes was **hermaphroditic**. Entolecithal eggs were produced in the female gonad. The sperm are thread-like, as in the Gnathostomulida, and have a cilium. They were probably transferred to the mate by means of a copulatory organ. The hypodermic injection results in internal fertilization of the egg cells. There were no oviducts; there were no shell glands. The eggs with a soft shell were probably pressed outwards through the skin.

Finally, spiral quartet cleavage and direct development without a larval stage belongs to the ground pattern. The stem species of the Plathelminthes could probably reproduce vegetatively by paratomy.

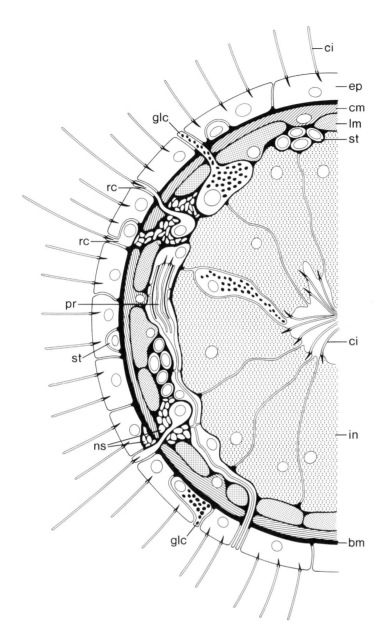

Fig. 60. Ground pattern of the Plathelminthes (Cross section of the body; without genital organs). ba = basement membrane and extracellular matrix. ci = cilium. cm = circular muscle. ep = multiciliary epidermis cell. glc = glandular cells in and beneath the epidermis. in = intestine with multiciliary cells and aciliated glandular cells as producers of digestive enzymes. lm = longitudinal muscle. ns = nerve cells located intra- and subepithelially. pr = protonephridium with two cilia in the terminal cell. rc = monociliary receptor cells in and beneath the epidermis. st = stem cells in and beneath the epidermis. (Ehlers 1995)

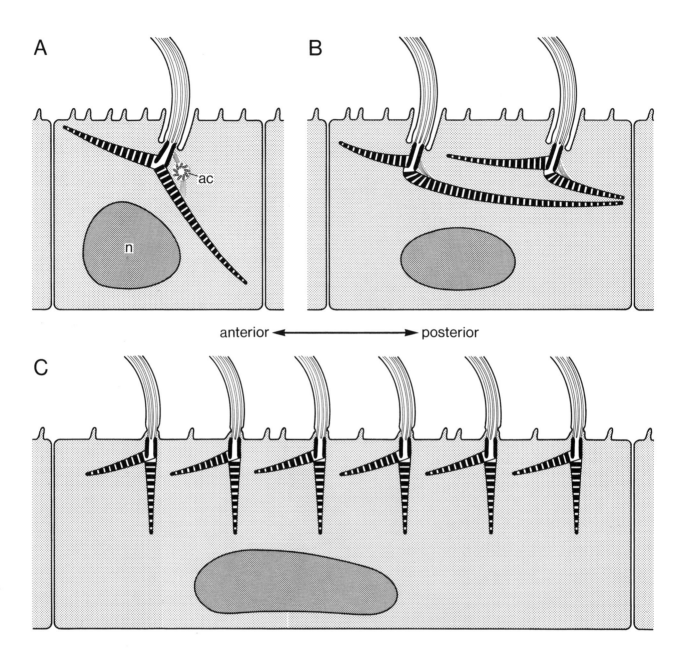

A

B

anterior ◄━━━━━━━━━━━━━► posterior

C

Fig. 61. Ciliation of the epidermis in the Plathelminthomorpha (longitudinal section). The cilia have two ciliary rootlets in the ground pattern – a rostral rootlet oriented towards the front and a caudal rootlet oriented towards the back or downwards. A. Gnathostomulida. Monociliary epidermis cells with accessory centriole next to the basal body. B and C. Plathelminthes. Multiciliary epidermis cells without an accessory centriole beside the basal body of the cilia. B. Catenulida. Weakly multiciliary epidermis cell. C. Euplathelminthes. Strongly multiciliary epidermis cell. ac = accessory centriole. n = nucleus. Dense of ciliation per μm^2 of skin surface: Gnathostomulida 0.15 – 0.2 cilia; Catenulida 0.2 – 1.8 cilia; Euplathelminthes three to six cilia. (Rieger and Mainitz; U. Ehlers, in Ax 1984)

Systematization

We will begin with a hierarchical tabulation of the phylogenetic system of the Plathelminthes. An alternative representation of the same contents is offered in three kinship diagrams which will be integrated in the course of the following text.

> **Plathelminthes**
>   **Catenulida**
>   **Euplathelminthes**
>   **Acoelomorpha**
>   **Nemertodermatida**
>   **Acoela**
>   **Rhabditophora**
>   **Macrostomida**
>   **Trepaxonemata**
>   **Polycladida**
>   **Neoophora**
>   **?Lecithoepitheliata**
>   **?Prolecithophora**
>   **Seriata**
>   **Rhabdocoela**
>   **"Typhloplanoida"**
>   **Doliopharyngiophora**
>   **"Dalyellioida"**
>   **Neodermata**
>   **Trematoda**
>   **Aspidobothrea**
>   **Digenea**
>   **Cercomeromorpha**
>   **Monogenea**
>   **Cestoda**

Catenulida – Euplathelminthes

In the first dichotomy of the Plathelminthes, the Catenulida are placed as the sister group opposite all other taxa that have been united under the name Euplathelminthes (Fig. 62).

With two cilia in the terminal cell of the protonephridium, the Catenulida exhibit the original number of cilia of the Plathelminthes. The epidermis cells, too, are only weakly multiciliated with around two cilia per cell. In contrast to the Euplathelminthes, the "frontal organ" is lacking, and this absence can also only be assessed as a plesiomorphy. On the other hand, the Catenulida have autapomorphous features. Two of these are the unpaired protonephridium, and the dorsal position of the male reproductive organ and the genital pore.

With regard to these last two features mentioned, the Euplathelminthes have the plesiomorphous alternatives, i.e., paired protonephridia and a ventral male pore. Autapomorphies in the Euplathelminthes are seen in the more highly multiciliated epidermal cells which have three to six cilia/μm^2 of epidermis (EHLERS 1986) and the frontal organ, which will be dealt with below.

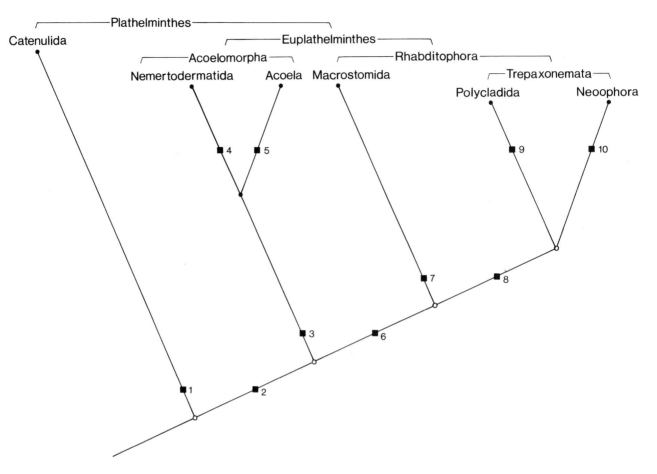

Fig. 62. Diagram of the phylogenetic kinship relations in the taxon Plathelminthes with the Catenulida and all other flat-worms – the Euplathelminthes – as sister groups. This diagram accompanies the text up to the Neoophora with the forma-tion of a heterocellular gonad (germovitellarium).

Catenulida

■ **Autapomorphies**
(Fig. 62 → 1)

– Unpaired dorsomedially positioned protonephridium with the nephropore at the posterior end of the body (Fig. 63A).
This can be incontestably interpreted as a derived feature in the out-group comparison with the paired protonephridia of the Gnathostomulida and in the in-group comparison with the Euplathelminthes.

– Ciliary rootlets as supporting elements in the terminal cells. A characteristic autapomorphy is associated with the original number of two cilia per terminal cell in the protonephridia. The two ciliary rootlets elongate into a rod-like shape, bend over after the origin from the basal body, forming an arch, and stretch out in a distal direction. In this manner, the ciliary rootlets form two supports on the narrow side of the weir of the terminal cell (Fig. 63B,C).

- The testes and male genital pore are located dorsally in the anterior section of the body (Fig. 63A). Based on this characteristic, the taxon name Notandropora (REISINGER 1924) was proposed. The name is highly appropriate because it is derived from an autapomorphy. However, the name Catenulida, which refers to the plesiomorphous formation of zooid chains by means of asexual reproduction, has remained in use.

- Aciliated, nonmobile sperm which form a transitory cilium only during spermiogenesis.

The Catenulida measures only a few millimeters. Well-known limnetic taxa of the traditional classification are the Catenulidae (*Catenula*), Stenostomidae (*Stenostomum, Rhynchoscolex*), and Chordariidae (*Chordarium*). A marine taxon, Retronectidae (*Retronectes, Paracatenula*), with numerous and widely distributed species was discovered only later (STERRER and RIEGER 1974). The marine Retronectidae are positioned as an adelphotaxon opposite all the limnetic Catenulida (EHLERS 1994a).

Apart from the autapomorphies just outlined, two further characteristic features of the Catenulida should be noted: the pharynx simplex and the widespread presence of a statocyst.

The pharynx simplex is a simple, ciliated tube between the oral pore and the ciliated intestinal cells. We shall describe its structure in species of the taxon *Retronectes* (DOE 1981; EHLERS 1985).

Sparsely ciliated epidermal cells surround the oral opening. In *Retronectes atypica*, interior to the oral opening is a transitory zone also composed of epidermal cells. Parts of the cells containing the nucleus, however, are deeply sunken in the body interior. Glands with circularly arranged ducts lead into the distal section of the pharynx. The following pharyngeal epithelium is clearly distinguishable from the epidermis. The individual epithelial cells are characterized by dense ciliation. The caudal ciliary rootlets are conspicuously elongated. The pharyngeal tube is surrounded by circular and longitudinal muscles (Fig. 64A, C).

A pharynx simplex with this construction is indisputably part of the ground pattern of the Catenulida. It is uncertain, however, whether this feature is a plesiomorphy or an autapomorphy of the Catenulida. Currently, a simple oral pore is hypothesized for the stem species of the Plathelminthes (p. 142). According to this opinion, ciliated pharyngeal tubes must have originated several times independently – in the Catenulida, sporadically within the Acoelomorpha and in the stem lineage of the Rhabditophora (p. 159). The pharynx simplex outlined above would, in other words, be a further autapomorphy of the Catenulida.

Let us look at the second feature. A statocyst is widespread within the Catenulida, but not always present; it is absent, e.g., in the limnetic species of *Stenostomum*. As opposed to all other Plathelminthes having a statocyst, the statocyst in the Catenulida is always behind the brain. A few parietal cells line the thin wall of the capsule. The statolith is found in the intracapsular cavity. Up to six statoliths can be present. No formative cells of the statolith have been discovered up to now (Fig. 64A).

In comparison with the gravity receptors of the Acoelomorpha (p. 154), considerable differences rule out a homology with the statocysts of the Nemertodermatida or of the Acoela. Moreover, there are doubts as to whether a statocyst is even part of the ground pattern of the Catenulida (EHLERS 1985, 1991).

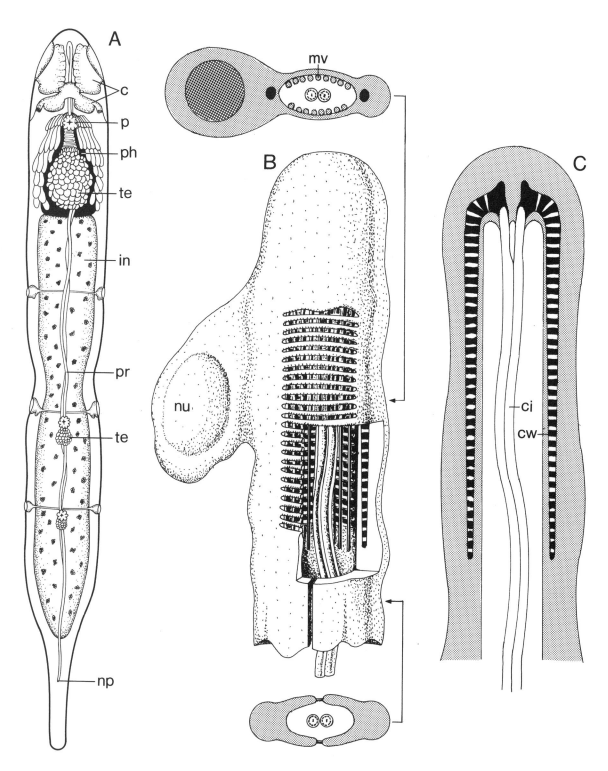

Fig. 63. Catenulida. Construction of the protonephridium. A. *Stenostomum sthenum* in asexual reproduction; a chain of several zooids. The unpaired protonephridium lies in the middle of the dorsal side. The nephropore is located in the posterior end. B. Ultrastructure of the biciliated terminal cell of limnetic Catenulida. C. Longitudinal section through the terminal cell to demonstrate the two median cilia and their rod-shaped extended ciliary rootlets in the narrow sides. c = brain. ci = cilium. cw = ciliary rootlet. in = intestine. mv = microvillus. np = nephropore. nu = nucleus. p = male pore. ph = pharynx. pr = protonephridium. te = testis. (A Borkott 1970; B and C Ehlers 1985, 1994a)

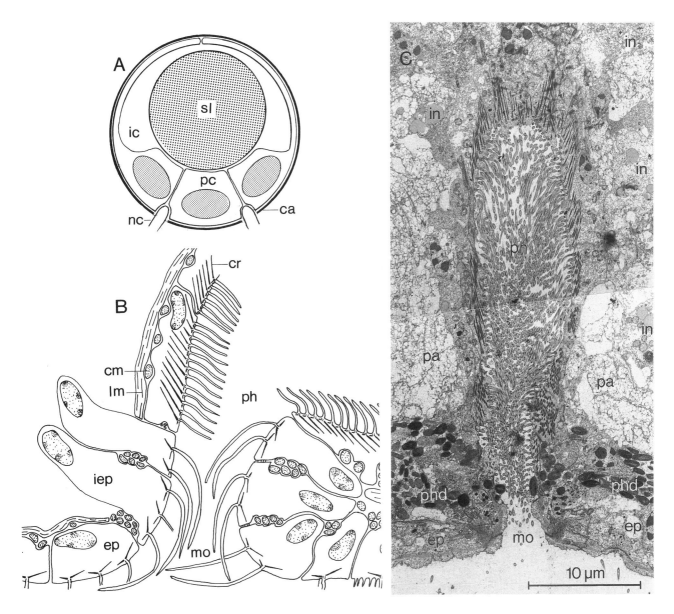

Fig. 64. Catenulida. A. Organization of the statocyst in the marine taxon Retronectidae. A statolith in an intracapsular cavity; three parietal cells in the periphery and two processes of extracapsular nerve cells. B. Entrance into the pharynx simplex of *Retronectes atypica* (Retronectidae). Following the normal epidermis is a transitory region with weakly ciliated and deeply sunken epidermis cells. The pharyngeal epithelium which follows thereafter consists of densely ciliated cells. C. Pharynx simplex of *Retronectes cf. sterreri* (Retronectidae) shown in a cross section of the body. Pharyngeal glands located close above the mouth. ca = capsule wall. cm = circular muscle. cr = caudal ciliary rootlet. ep = epithelium. ic = intracapsular cavity. iep = sunken epidermis cell. in = intestine. lm = longitudinal muscle. mo = mouth. nc = nerve cell. pa = parenchym. pc = parietal cell. ph = pharynx. phd = pharyngeal gland. sl = statolith. (A Ehlers 1991; B Doe 1981; C Ehlers, in Ax 1984)

Euplathelminthes

■ **Autapomorphies**
(Fig. 62 → 2)

– Densely multiciliated epidermal cells.

– Frontal organ or frontal glandular complex.
Accumulation of different glandular cells and ciliated sensory cells in the anterior tip of the body. In the ground pattern of the Euplathelminthes, the glands do not lead outwards through a common apical pore. The term frontal organ is thus only used with reservations (EHLERS 1992c).

Acoelomorpha – Rhabditophora

The Acoela with a syncytial digestive tissue are probably the best-known taxon of free-living Plathelminthes – at least in phylogenetic speculation on the "origin" of the Bilateria, and even of the Metazoa as a whole.

The seemingly simple organization of the Acoela, in which the gut lumen is absent, was used to trace the Bilateria back to compact planula larvae of the Cnidaria that also have no gut lumen. Due to the syncytial character of the intestinal tissue, the Acoela were even placed at the "basis" of all Metazoa, deriving them from multinucleate, ciliate-like unicellular organisms.

Today, ideas of this kind are of historical interest only; they represent an outdated phase of unprovable speculation on the phylogenetic development of the Metazoa.

This can be easily demonstrated. The Acoela are at the "basis" neither of the Bilateria or of the Metazoa, nor are they the "most original" taxon of the Plathelminthes. Rather, together with the sister group Nemertodermatida, they form a monophyletic subtaxon Acoelomorpha within the Euplathelminthes that has a series of apomorphous features. The syncytial, glandless digestive tissue of the Acoela represents an extremely derived state. Today, its evolution can be documented step by step.

The Rhabditophora form the adelphotaxon of the Acoelomorpha. The original, ciliated intestinal epithelium with glandular cells has been retained in their ground pattern. On the other hand, characteristic features of the Rhabditophora are the lamellate rhabdites, which are a special form of solid, rod-shaped secretions of the skin, and the duo-gland adhesive organ.

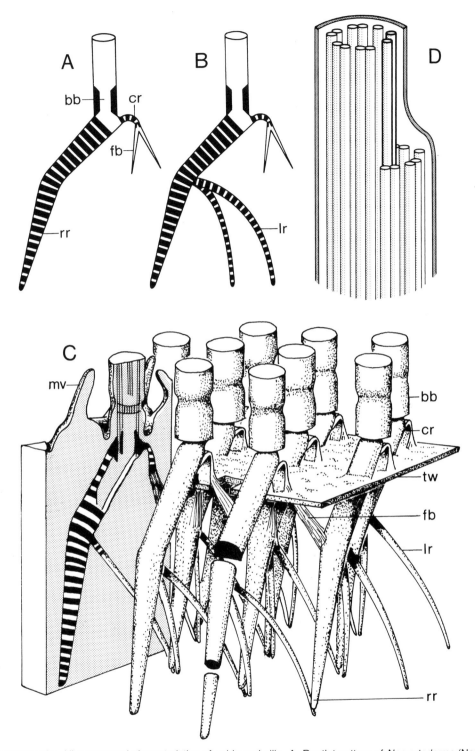

Fig. 65. Acoelomorpha. Ultrastructural characteristics of epidermal cilia. A. Rootlet pattern of *Nemertoderma* (Nemertodermatida). Large rostral rootlet with knee; two fiber bundles arise from the small caudal rootlet. B. Rootlet pattern of the Acoela. Two additional lateral rootlets arise from the knee area as an evolutionary novelty. C. *Chilia groenlandica*. Rootlet pattern of the cilia of the Acoela in a three-dimensional representation. D. Distal end of the epidermal cilia in the Acoelomorpha. Four peripheral double tubules end partially underneath the cilia tip. bb = basal body. cr = caudal ciliary rootlet. fb = fiber bundle. lr = lateral rootlet. mv = microvillus. rr = rostral ciliary rootlet. tw = terminal web. (A, B, D Ehlers, in Ax 1984. C Hendelberg and Hedlung 1974)

Acoelomorpha

The Acoelomorpha include the sister groups Nemertodermatida and Acoela. First, we will establish the obvious autapomorphies of the three taxa Acoelomorpha, Nemertodermatida, and Acoela, and then trace the evolution of the special features of the epidermis and the digestive systems.

■ **Autapomorphies**
(Fig. 62 → 3)

– Network formed by interconnecting rootlets of epidermal cilia.
Each locomotory cilium has two rootlets in the ground pattern of the Plathelminthes. One rootlet runs horizontally towards the front, the other is slanted towards the back or downwards. In the ground pattern of the Acoelomorpha, the large rostral rootlet is angled in the middle. The smaller, caudal rootlet is divided into two laterally radiating bundles of fibers (Fig. 65A). These bundles are connected with the two rostral rootlets positioned to the left and right behind them where they are angled. The linking of neighboring ciliary rootlets forms an interconnected root system peculiar to the Acoelomorpha.

– Shaft region in epidermal cilia.
In the locomotory cilia, four of the nine peripheral double tubules (Nos. 4–7) end a considerable distance below the tip of the cilium. Only five peripheral units, usually in the form of double tubules, appear in the reduced shaft along with the two central tubules (Fig. 65D).

– Absence of protonephridia.
All Nemertodermatida and Acoela lack a protonephridial system. This is a peculiar phenomenon within the Plathelminthomorpha. Due to their presence in the Gnathostomulida, the Catenulida, and in other unities of the Euplathelminthes, we can justifiably postulate that the protonephridia are part of the ground pattern of the Plathelminthes. The lack of protonephridia in the Acoelomorpha must be interpreted as an apomorphous state.

Fig. 66. Representatives of the Acoelomorpha (upper row) and evolution of the digestive tissue within the descent community (lower row). A. *Nemertoderma* (Nemertodermatida). Dorsal view of a living organism. B. *Nemertoderma.* Diagram showing organization in a sagittal section. C. *Paratomella unichaeta* (Acoela) with asexual reproduction through paratomy. Representative of the taxon *Paratomella*, which is interpreted as the sister group of all other Acoela – Euacoela. D. *Mecynostomum auritum* (Euacoela). E-G. Cross sectional representations. E. *Nemertoderma.* Original state of the digestive system with a single-layered intestinal epithelium composed of digestive cells and enzyme producing glandular cells. F. *Paratomella.* Double-layered intestinal tissue composed of central and peripheral cells; glandular cells are reduced. G. ▶

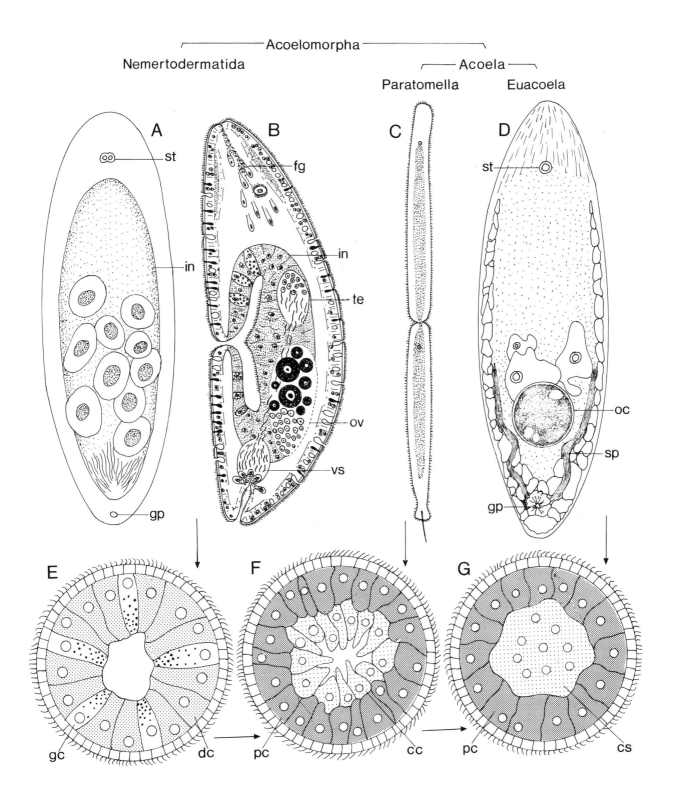

Euacoela. Central cells fuse together to form a syncytium without a lumen. cc = central intestinal cell. cs = central syncytium. dc = digestive cell. fg = frontal glands. gc = glandular cells of the intestinal epithelium. gp = male genital pore. in = intestine. oc = oocyte. ov = ovary. pc = peripheral intestinal cell. sp = spermal cord. st = statocyst. te = testis. vs = vesicula seminalis. (A Westblad 1937; B Karling 1974; C Ax and Schulz 1959; D Luther 1960. E-F drawings following Smith III and Tyler 1985 under assessment of information in Ehlers 1992a; kinship diagram at the top Ehlers 1992a)

Nemertodermatida – Acoela

Nemertodermatida

■ **Autapomorphy**
(Fig. 62 → 4)

– Statocyst with two statoliths and several parietal cells.
A series of parietal cells are found in the periphery of the statocyst. A plasma process from one cell divides the statocyst into two chambers. A statolith is produced in each chamber by a statolith-producing cell (lithocyte; Fig. 67B).

Acoela

■ **Autapomorphies**
(Fig. 62 → 5)

– Further differentiation of the root system of epidermal cilia by additional lateral rootlets.
Two new lateral rootlets originate where the large rostral root is angled. They run obliquely towards the back where they touch the tips of the left and right rostral rootlets positioned behind them. This results in an additional interlinking of the ciliary rootlets in the epidermis (Fig. 65B,C).

– Statocyst with one statolith and two parietal cells.
The two parietal cells line the statocyst; their nuclei are positioned dorsolaterally. The lithocyte lies in a voluminous intercellular space and produces a lenticular statolith (Fig. 67C). A homology between the statocysts of the Nemertodermatida and Acoela is unlikely. For this reason, the different manifestations of the statocyst are interpreted to be two autapomorphies which evolved independently of each other in the stem lineages of the Nemertodermatida and Acoela.

– Complete absence of an extracellular matrix. Using the out-group comparison this, too, as in the case of the protonephridia of the Acoelomorpha, must be interpreted as a derived state.

– Digestive system without glandular cells.

– Sperm with two cilia enclosed in the body of the sperm.
The original state of monociliated sperm is realized in the adelphotaxon Nemertodermatida.

– Spiral duet cleavage.
In contrast to the plesiomorphous spiral quartet cleavage, the Acoela are characterized by the production of micromere duets at the animal pole. Compared with the four stem blastomeres A, B, C, and D of the quartet cleavage, the Acoela have only two stem blastomeres,

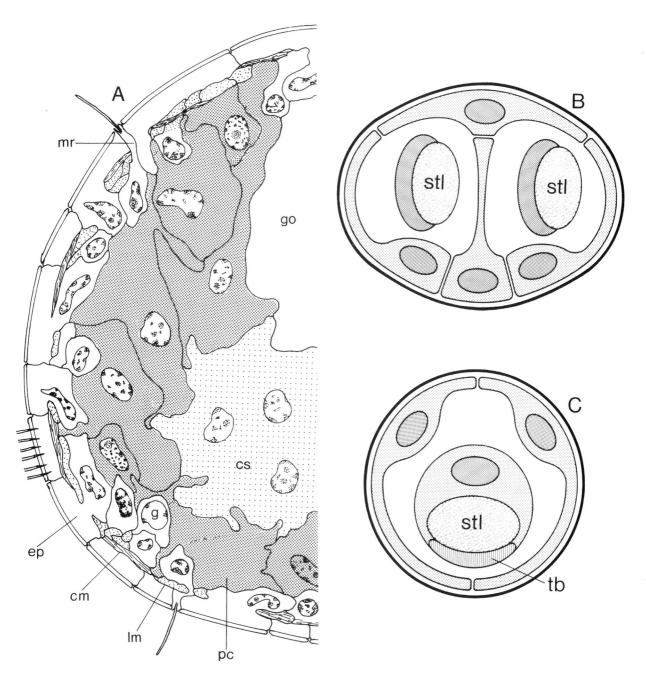

Fig. 67. Acoelomorpha. A. *Diopisthoporus cf. longitubus* (Acoela). Representation of tissue organization in a cross section of the body. The cellular epidermis (light, cilia shown in only one cell) is followed first by the peripheral cellular intestinal tissue (darkly dotted), and then by the central digestive syncytium (lightly dotted). Cells of the gonad region are not shown. B. Statocyst of *Nemertoderma* (Nemertodermatida) with two statoliths and several parietal cells (Cross section). C. Statocyst of the Acoela with one statolith and two parietal cells (Cross section). In the Euacoela, the statolith building cell produces a special tubular body beneath the statolith. cm = circular muscle. cs = central syncytium. ep = epidermis cell. g = epidermal glandular cell. go = gonad region. lm = longitudinal muscle. mr = monociliary receptor cell. pc = peripheral intestinal cell. stl = statolith. tb = tubular structure. (A Smith III and Tyler 1985; B and C Ehlers 1985)

The structure of the E p i d e r m a l S u p p o r t S y s t e m which is composed of ciliary rootlets is documented in two stages (Fig. 65). In the Nemertodermatida, two fiber bundles of the caudal rootlet connect with neighboring ciliary rootlets. This has already been described under their autapomorphies. In the Acoela, new lateral rootlets appear in the next evolutionary step, which serve as additional intraepidermal strengthening. A unique support system composed of ciliary rootlets, which is unknown in other Metazoa, evolved here. This is evidently due to the extensive or complete lack of an ECM (EHLERS 1992a). A very thin granular matrix can "still" be seen in different regions of the body of the Nemertodermatida, but the ECM under the epidermis is absent. All Acoela are characterized by the complete lack of an extracellular matrix. The reduction of a basal lamina beneath the epidermis must be seen in relation to the evolution of a ciliary support system in the epidermis.

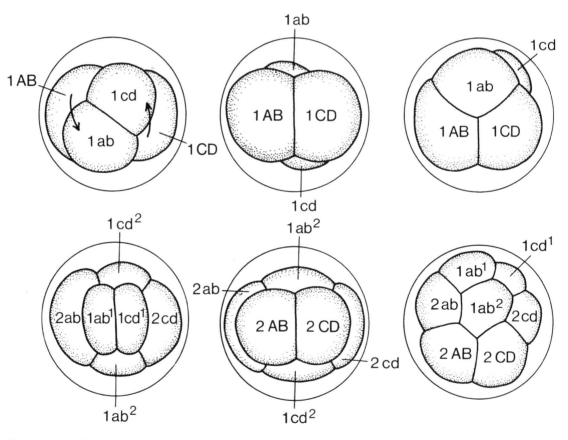

Fig. 68. Acoela. Spiral duet cleavage in *Oligochoerus limnophilus*. Top row: four cell stage. From left to right: view of the animal pole, vegetative pole, and view from the side. Lower row: eight cell stage arranged accordingly. The two stem blastomeres (AB, CD) in the duet cleavage correspond to the four stem blastomeres (A, B, C, D) in the quartet cleavage. (Ax and Dörjes 1966)

Let us examine the Digestive System of the Acoelomorpha. As before, we can begin with the Nemertodermatida, whose structure differs only slightly from the ground pattern of the Plathelminthes. The Nemertodermatida (Fig. 66A,B,E) have a single-layered intestinal epithelium comprising digestive cells and glandular cells. A gut lumen may be present or absent. The epithelial structure of the digestive system no longer exists amongst the Acoela. The glandular cells have been completely reduced. The structure of *Paratomella* (Fig. 66C,F) is very informative. A purely cellular digestive system with a central lumen has been retained here. A study of the ultrastructure reveals that the central and peripheral cells form two layers of intestinal tissue.

In the stem lineage of the Euacoela, the central cells fused together to form a periodic or permanent syncytial tissue leading to the evolution of the digestive syncytium without a gut lumen (Figs. 66D,G, 67A). In the periphery, however, the cellular character of the digestive tissue has been retained. A long and intensely disputed problem as to the histological structure of the Acoela was thus solved by a comparative analysis of the ultrastructure.

It remains to be said that the unfortunate term "digestive parenchyma" should be avoided when referring to the Acoela tissue described above. This tissue is not a parenchyma between the epidermis and intestine, but rather the endodermal intestinal tissue itself.

Rhabditophora

■ **Autapomorphies**
(Fig. 62 → 6)

– Increase in the number of cilia in the terminal cells of the protonephridia.
Compared to the two cilia of the Catenulida, the number of cilia here have doubled to four; this is seen in the Macrostomida. The further evolutionary development that took place within the Euplathelminthes led to terminal cells with numerous cilia.

– Lamellate rhabdites.
Free-living Plathelminthes produce a wide range of solid secretions (rhabdoids) in the epidermis or in subepidermal glands. Today, the term rhabdites (lamellate rhabdites) is used only for rhabdoids which have a cortical layer (cortex) composed of concentrically arranged lamella, and a medulla which varies from homogenous to granular (Fig. 69 B). Lamellate rhabdites with this specific ultrastructure are not found in the Catenulida and Acoelomorpha. The presence of lamellar rhabdites in the Macrostomida, Polycladida, and in the ground pattern of the Neoophora, is an outstanding characteristic feature of the taxon Rhabditophora in which they are included.
Under these circumstances, the lack of rhabdites amongst a series of smaller subtaxa of the Rhabditophora and the parasitic Neodermata (trematodes and tapeworms) must be interpreted as the result of evolutionary reduction.

– Duo-gland adhesive organ.
The Rhabditophora have found an elegant solution to the vital problem of how to anchor themselves quickly and free themselves just as quickly from grains of sand or other particles in an agitated benthic substrate.

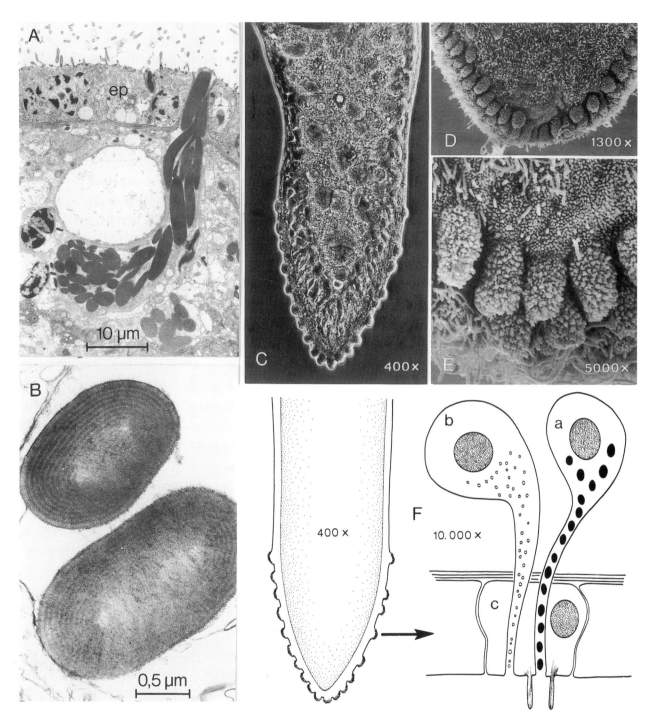

Fig. 69. Rhabditophora. A and B. Rhabdites and C-F. Duo-gland adhesive organ as autapomorphies of the descent community. A. *Cirrifera aculeata* (Seriata). Production of rod-shaped rhabdites in glandular cells beneath the epidermis (ep). B. *Notoplana atomata* (Polycladida). Cross sections of rhabdites with concentric lamella in the periphery. C. *Bothriomolus balticus* (Seriata), as well as D and E. *Coelogynopora axi* (Seriata). Posterior ends with adhesive papillae. F. Diagram of an adhesive organ from the papillae. The organ consists of a glandular cell (a) with large secretions, a second glandular cell (b) with small grana and an epidermal anchor cell (c). (A, D-F Sopott-Ehlers, in Ax 1984; B Ehlers 1985)

In the stem lineage of the Rhabditophora, a tiny adhesive organ comprising three kinds of differentiated cells evolved (TYLER 1976). These are:

1. a glandular cell which secretes large electron dense grana that causes adhesion to the substrate;

2. a second glandular cell with much smaller grana which reverse the adhesion, probably with the help of lytic properties;

3. an anchor cell which is a modified epidermal cell that holds both glandular cells closely together. The ducts of the glands pass through the anchor cell close to each other. A microvilli collar surrounding the pore of the adhesive cell serves to spread the secretion onto the substrate (Fig. 69C-F).

Even a system as effective as this disappears again in the course of evolution. Within the Rhabditophora, the duo-gland adhesive organ is absent in the Doliopharyngiophora, including the trematodes and tapeworms mentioned above.

– Pharynx simplex coronatus (Fig. 71).
The simple pharyngeal tube of the Macrostomida is probably not homologous to the pharynx simplex of the Catenulida, although there are obvious congruences: epidermal cells around the oral opening, transitory zones with rings of sunken epidermal cells, a ring of glandular cells (with different secretions) on the border between the transitory zone and the pharyngeal epithelium proper, the latter having again long caudal ciliary rootlets.

A new element is a thick nerve ring (nr) around the lower part of the pharyngeal tube which led to the term pharynx simplex coronatus. This nerve ring also appears in strongly differentiated pharynges of the Rhabditophora and can therefore serve as an autapomorphy of this unity (EHLERS 1985).

The pharynx simplex coronatus is found only in species of the taxon Macrostomida. It must, however, be considered as part of the ground pattern of the Rhabditophora, because more evolved pharynges, known under the collective term pharynx compositus, can be traced back to the pharynx simplex coronatus.

Macrostomida – Trepaxonemata

Compared with the adelphotaxon, the Macrostomida are characterized by a large number of original features. Let us therefore begin with this unity.

Macrostomida

■ **Autapomorphies**
(Fig. 62 → 7)

– Post oral nerve commissure.
The nervous system has two main ventrolateral nerve cords. There is a prominent connection between them immediately behind the mouth. This is a marked characteristic of the Macrostomida (po in Figs. 70C,71).

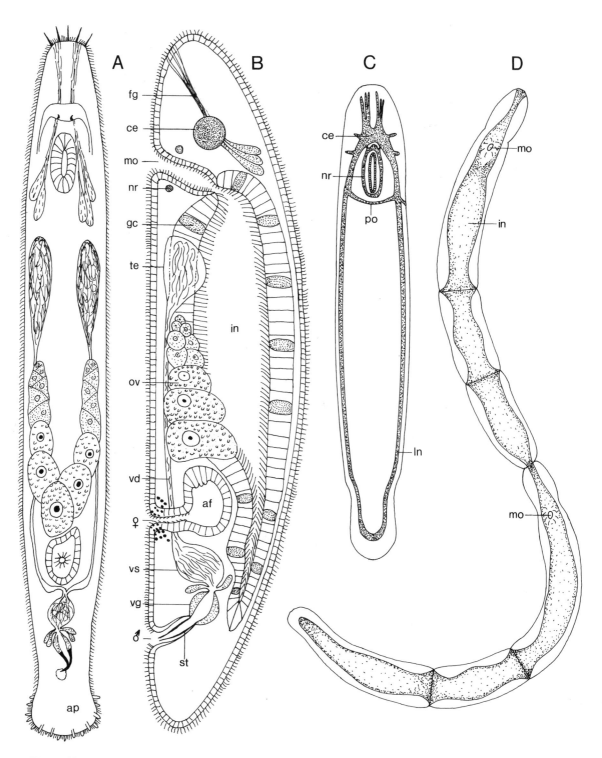

Fig. 70. Macrostomida. A. *Macrostomum*. Dorsal view of organization. B. *Macrostomum*. Sagittal section. C. *Macrostomum*. Nervous system with nerve ring around the pharynx simplex (autapomorphy of the Rhabditophora) and postoral nerve commissure (autapomorphy of the Macrostomida). C. *Microstomum jenseni* in asexual reproduction. Animal chain with several zooids. af = antrum femininum. ap = adhesive plate with papillae into which run the glands of the adhesive organs. ce = brain. fg = frontal glands. gc = gland of the intestinal epithelium. in = intestine. ln = longitudinal nerve. mo = mouth. nr = nerve ring. ov = ovary. po = postoral nerve commissure. st = stylet of the copulatory organ. te = testis. vd = vas deferens. vg = vesicula granulorum. vs = vesicula seminalis. (A Ax 1966b; B Ax 1963; C Luther 1905; D Faubel 1974)

– Aciliated sperm.
Electron microscope studies of species from different taxa have shown that sperm of the Macrostomida are aciliated. This is without doubt an apomorphous state compared with the widespread existence of cilia in the sperm of the Plathelminthes.

As is the case in all the unities of the Plathelminthes which have been dealt with up to now, the Macrostomida are only a few millimeters in size. *Microstomum* with the original vegetative form of reproduction by means of paratomy and *Macrostomum* are common taxa, with species found throughout the world in oceans and fresh water.

Let us emphasize a representative selection of original characteristics of the Macrostomida within the Rhabditophora: pharynx simplex coronatus, simple sack-shaped intestine with ciliated digestive cells, paired testes, and paired ovaries with entolecithal egg production, male and female genital apertures primarily separated from each other (Fig. 70).

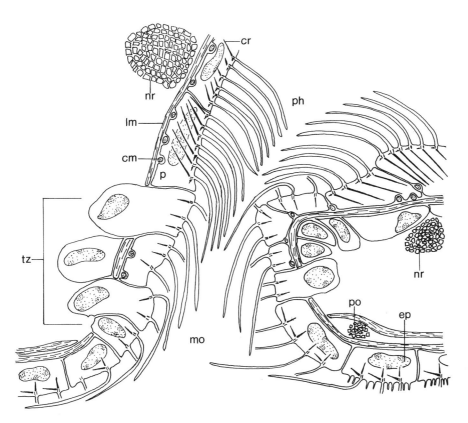

Fig. 71. Macrostomida. Pharynx simplex coronatus in *Paramyozonaria simplex*. Lower pharyngeal part with nerve ring and postoral nerve commissure shown in a sagittal section; based on studies of the ultrastructure. cm = circular muscle. cr = caudal ciliary rootlet. ep = epidermis cell. lm = longitudinal muscle. mo = mouth. nr = nerve ring. p = pharyngeal cell. ph = pharynx. po = postoral commissure. tz = transitory zone between epidermis and pharyngeal epithelium. (Doe 1981)

Trepaxonemata

(Fig. 62 → 8)

– Biciliated sperm.
The spermatozoa have two cilia. As against the monociliated sperm in the ground pattern of the Plathelminthes, found in the Nemertodermatida (p. 133), this is an apomorphy. The two cilia protrude freely from the sperm. Amongst the Neodermata, the inclusion of the cilia in the cytoplasm of the sperm (p. 181) is a derived state.

– Central axial rod in the cilia of the sperm (Fig. 72).
Sperm whose cilia have two central microtubules and nine peripheral double tubules are part of the ground pattern of the Plathelminthes. This plesiomorphous 9 x 2 + 2 pattern has been established for the Acoelomorpha. Among all Plathelminthes united under the name Trepaxonemata, the two central tubules are replaced by an unpaired axial rod (9 + 1 pattern).
The rod is composed of: (1) a central electron dense element, (2) an intermediate light zone, and (3) a peripheral, electron-dense sheath with longitudinal bands running spirally. Nine spokes connect the sheath with the nine peripheral double tubules (THOMAS 1975).

– Pharynx compositus
In contrast to the pharynx simplex coronatus of the Macrostomida, all Trepaxonemata have complex, highly muscular pharynges which can protrude far out of the oral opening. They are united under the name pharynx compositus. This does not, of course, say anything about a possible homology between different manifestations of the pharynx compositus. This question has not yet been resolved, and so the feature "pharynx compositus" can be added to the list of autapomorphies only with reservations. However, a pharynx plicatus definitely represents a plesiomorphous state of the pharynx compositus; it is found in the Polycladida, the Seriata, and parts of the Prolecithophora.
Basic characteristics of the pharynx plicatus are the pharyngeal fold with an open connection to the body tissue, and the pharyngeal pocket, which completely surrounds the pharynx. The evolution of the pharynx plicatus may have been the result of an inward fold from the region of the mouth of the pharynx simplex coronatus. As a result, the nerve ring surrounding the pharynx simplex coronatus was included in the body of the pharynx plicatus.
The oral opening of the pharynx simplex coronatus is located on the ventral side; the pharyngeal tube runs obliquely or vertically into the body. Thus, for the pharynx plicatus a ventral mouth and its alignment perpendicular to the longitudinal axis of the body may form the original state.

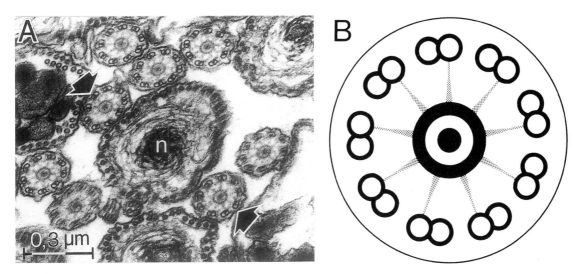

Fig. 72. Trepaxonemata. Ultrastructure of the cilia of the sperm. A. *Notocaryoplana arctica* (Seriata). A sperm with nucleus (n) is located in the center of the cross section. To the left and right are the two cilia belonging to the sperm marked by arrows. B. Cross-sectional diagram of a sperm cilium. A central axial rod is surrounded by nine peripheral double microtubules. The unpaired axial rod forms an autapomorphy of the Trepaxonemata. (Ehlers, in Ax 1984)

Polycladida – Neoophora

The Polycladida are the only taxon within the Trepaxonemata with a homocellular ovary exhibiting the original method of entolecithal egg production. A heterocellular, female gonad with germocytes and vitellocytes as well as the formation of ectolecithal eggs evolved in the stem lineage of the sister group, Neoophora. We will deal with these phenomena in greater detail, but first we must justify the taxon Polycladida as a monophylum.

Polycladida

■ **Autapomorphies**
(Fig. 62 → 9)

– Increase from millimeter sized body to centimeters.
  The following phenomena are related to this increase:

– Dorsoventral flattening. The Polycladida are the first "real" flatworms.

– Evolution of numerous intestinal diverticula towards the front, sides, and back of the body; in the periphery, the diverticula may fuse together to form a network. A highly differentiated gastrovascular system originated in the stem lineage of the Polycladida from the primarily simple sack-shaped intestine of the Plathelminthes (Fig. 73).

– An extensive parenchyma fills the spaces between the intestinal diverticula.

– Division of the testes and ovaries into numerous follicles, whereby the division of the ovary in particular is unique within the Plathelminthes.

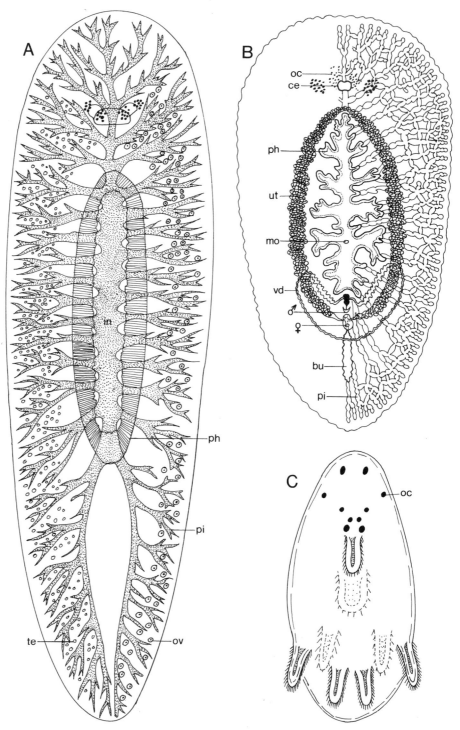

Fig. 73. Polycladida. A and B. Structure of the intestinum with formation of diverticula in the entire body. A. *Notoplana alcinoi*. The intestinal diverticula originate above the pharynx and branch out in the periphery. (The only genital organs shown here are the testes follicles on one side and the ovarian follicles on the other side). B. *Leptoplana timida*. Net-shaped fusion of the intestinal diverticula in the periphery (testes and ovaries are not shown). C. Lobophora larva of the taxon Cotylea with eight lobe-shaped processes. Two lobes originate dorsally and ventrally in the middle body area; six ciliated lobes form a wreath around the posterior end. bu = bursa. ce = brain. in = intestine. mo = mouth. oc = ocellus. ov = ovarian follicle. ph = pharynx. pi = peripheral branching of the intestinal diverticula. te = testes follicle. ut = uterus. vd = vas deferens. (A de Beauchamp 1961b; B Bresslau 1928-33; C original – Bight of Kiel)

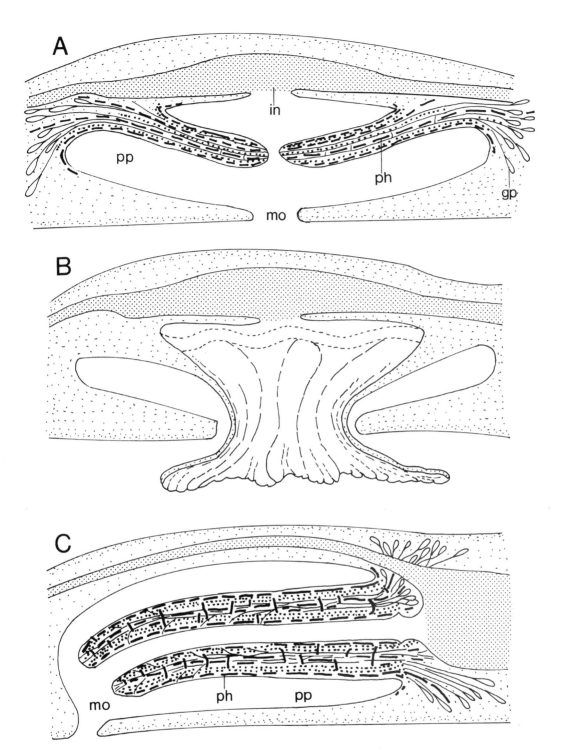

Fig. 74. Polycladida. States of the pharynx plicatus. Sagittal schemes. A and B. Ruffled pharynx with the original perpendicular position at the ventral side. A. Resting position of the pharynx. B. Outfolding from the mouth to take in food. C. Derived state of the cylinder pharynx of the Euryleptoidea. The tube lies horizontally in the body and is oriented towards the front. gp = pharyngeal glands. in = intestine. mo = mouth. ph = pharynx. pp = pharyngeal pocket. (de Beauchamp 1961b)

The traditional division of the Polycladida into the **Acotylea** and **Cotylea** (Lang 1884) is a classical example of typological classifications of taxa into paraphyletic and monophyletic subgroups – comparable with the now eliminated division of the Insecta into the Apterygota and Pterygota.

Just as the primarily wingless insects were brought together as "Apterygota" based on the lack of a feature, so, too, are the "Acotylea" based on the primary absence of a sucker behind the female genital aperture. The "Acotylea" form a paraphylum (Faubel 1983, 1984).

On the other hand, the presence of this glandulo-muscular adhesive organ among the Cotylea corresponds to the existence of wings in the Pterygota. This is undoubtedly an apomorphous feature. Let me take the comparison even further. Just as we base the hypothesis of a single evolution of wings in the stem lineage of the Pterygota on the most parsimonious explanation, the sukker can also be interpreted as a feature which originated once within the Polycladida and which can thus, as an autapomorphy, justify a monophylum Cotylea. It should be noted that, for the rest, a phylogenetic systematization of the entire Polycladida does not as yet exist.

Let us take a brief look at two phenomena found among the Polycladida – the manifestation of the pharynx plicatus and the existence of plankton larvae.

The Pharynx plicatus appears in two ground forms in the Polycladida (Fig. 74). A ruffled pharynx with a vertical position in the body and a ventral oral opening is common in the "Acotylea" and seen in many Cotylea. A cylindrical tube directed towards the front exists in only one subgroup of the Cotylea (Euryleptoidea). As a result of this distribution and assuming a primary ventral orientation of the pharynx plicatus (p. 162), we come to the following conclusion: the ruffled pharynx is the plesiomorphous state among the Polycladida and the cylindrical pharynx is an apomorphy within the taxon (Bock 1913, Faubel 1984).

We must take this argument one step further. The highly ruffled shape of the pharyngeal fold is found only among the Polycladida, which leads to the hypothesis that the ruffled shape originated as an evolutionary novelty in the stem lineage of the Polycladida. The ruffled pharynx as a special form of the pharynx plicatus is, in other words, probably another autapomorphy in the ground pattern of the Polycladida (Ehlers 1985).

Most of the "Acotylea" which do not have the adhesive organ develop directly. Only in single species of the Acotylea and in the Cotylea is there a free-swimming larva with four or eight (up to ten) lobe-shaped outgrowths (Fig. 73C). The backward-pointing processes of these Lobophora bear long cilia.

In phylogenetic discussions, the lobophora larvae (Mueller's larva, Goette's larva) are related to the trochophora larvae of the Trochozoa, and are therefore also considered the precursors (Protrochula) of the Trochophora (Jägersten 1972, Ruppert 1978, v. Salivini-Plawen 1980).

A homologous correspondence between the lobophore and trochophore larvae can, however, be definitely ruled out. The limitation to parts of the Polycladida indicates that the evolution of lobophora larvae first originated within the Polycladida (Ehlers 1985).

A lobophora does not belong to the ground pattern of the Polycladida, and is thus logically not part of the ground pattern of the Plathelminthes. A comparison with the trochophora in terms of a common evolutionary origin would be legitimate only if a free-swimming larval stage could be postulated for the life cycle of the stem species of all Plathelminthes. This, however, is not the case.

Neoophora

■ **Autapomorphies**
(Fig. 62 → 10)

– Germovitellarium = heterocellular female gonad.
Production of germocytes as germ cells capable of development and vitellocytes as nutritive cells incapable of development.

– Ectolecithal, compound "eggs". One germocyte and several vitellocytes are combined in an egg capsule.

The taxa of the Plathelminthes that have been dealt with up to this point have a homocellular female gonad. The ovary produces only one kind of cell, the entolecithal eggs. Drops of yolk develop as nutritive material for the embryos in the young oocyte, and among the Euplathelminthes, shell-drops which serve in the formation of an eggshell also develop. A homocellular ovary and entolecithal egg production are plesiomorphous features in the ground pattern of the Plathelminthes.

The Germovitellarium – a heterocellular female gonad – then developed in the stem lineage of the Neoophora. Two different female cells are formed. These are the Germocytes, from which the embryo develops, and the Vitellocytes, which are incapable of development containing nutritive material (drops of yolk, drops of fat, glycogen particles) and drops of shell-building material. As a result of this differentiation of the female gonad, Ectolecithal Eggs are produced. One germocyte (or also a few germ cells) and numerous vitellocytes are united into a compound "egg" and enclosed in an egg capsule (Fig. 75). While in the case of entolecithal egg formation the embryo is supplied with nutritive substances from the blastomeres, in the ectolecithal egg it takes up nourishment from the disintegrating yolk cells. These new demands led to the evolution of a special embryonic pharynx in the Tricladida (Fig. 80C).

A germovitellarium in the form of a mixed gonad, in which germocytes and vitellocytes are formed next to each other, must be regarded as a plesiomorphous state of the Neoophora. These conditions are found in the Lecithoepitheliata and the Prolecithophora, two taxa whose kinship within the Neoophora is still unclear.

The germovitellarium of the Lecithoepitheliata consists of a proximal zone for the production of germ and yolk cells and a distal part, in which each young germocyte is surrounded by a layer of vitellocytes. This epithelial-like arrangement of the yolk cells led to the name Lecithoepitheliata (Fig. 77A).

In the taxon Prolecithophora, the gonad in *Archimonotresis limophila* is said to be a highly original state. A single gonad, in which sperm, germ cells and yolk cells originate, forms a compact layer around the intestine. Spermatocytes and germocytes are found in the anterior dorsal area; the vitellocytes occupy the rest of this gonadal layer (Fig. 77B).

The zones in which germocytes and vitellocytes are formed are spatially separated in the majority of the Neoophora. The primarily uniform germovitellum is divided into two separate regions known as Germarium and Vitellarium. This division is an apomorphous state within the Neoophora.

All cases that have been analyzed in detail reveal that two different components participate in the formation of the Egg Shells or Egg Capsules. In addition to the drops of shell substances from the oocytes of entolecithal eggs or from the vitellocytes of ectolecithal eggs, there are secretions from special cells of the female genital tract. In the Neoophora, the secretions usually originate in localized shell glands which surround certain expandable sections of the efferent genital system. Among the parasites, these are the Mehlis's glands and the ootype. When single germocytes enter this area together with several vitellocytes, the shell glands produce a soft secretion which surrounds the egg and yolk cells. Drops of shell substance are released from the vitellocytes; they liquefy and fuse together with the secretion from the shell gland to form the egg shell (Fig. 75).

This egg shell can have a highly complex structure. Psammobiont Plathelminthes have developed pressure absorbing protective cushions for the egg. To illustrate this, I have selected two Kalyptorhynchia (Rhabdocoela) which are equipped with a proboscis. In *Diascorhynchus rubrus* the eggshell consists of a 7–9-µm-thick layer; a regular hexagonal system of braces creates a pattern which looks like a perfect copy of a honeycomb.

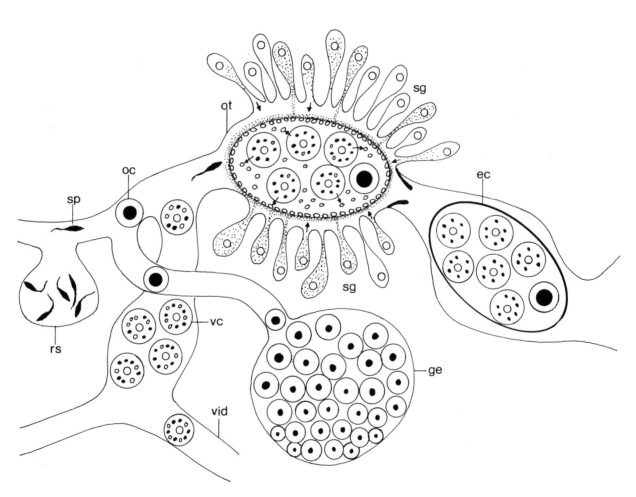

Fig. 75. Neoophora. The principle of the formation of compound, ectolecithal eggs, shown for a representative of the Digenea (Trematoda). Germocytes (oocytes) from the germarium and vitellocytes from the vitellaria will be brought together in an extensible part of the female genital apparatus (ootyp). Here, shell glands produce a liquid secretion which is put into the ootype from the outside. At the same time, inside, the vitellocytes release drops of shell substance. A hard egg shell (egg capsule) originates from the fusion of these two components. It usually surrounds one oocyte and several vitellocytes, which possess only drops of yolk at this point. The ectolecithal egg is finished. ec = egg capsule. ge = germarium. oc = germocyte (oocyte). ot = ootype. rs = receptaculum seminis. sg = shell glands. sp = sperm. vc = vitellocyte with drops of shell substance (light) and drops of yolk (black). vid = vitelloduct. (After Dönges 1988)

168

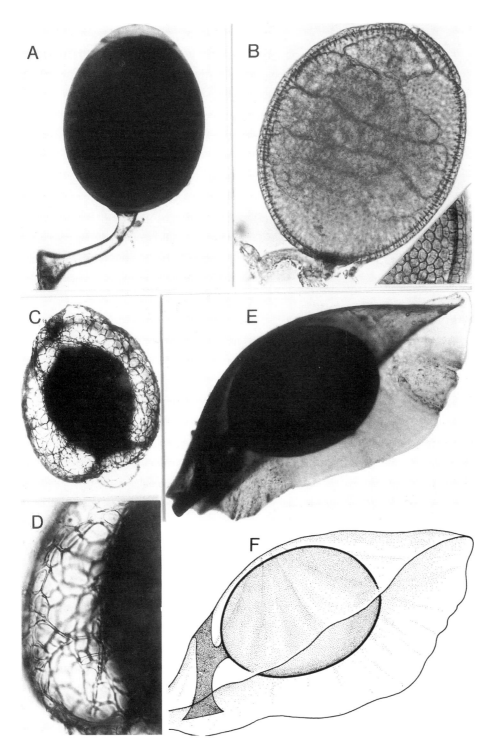

Fig. 76. Egg capsules of psammobiontic Neoophora with protective cushions and egg stalks. A. *Itaspiella helgolandica* (Otoplanidae, Seriata) with a long, elastic stalk composed of cement gland secretions. B. *Diascorhynchus rubrus* (Kalyptorhynchia). Protective covering of shell gland secretions which is attached to the egg shell in a honeycomb pattern (section below, right). C. *Diascorhynchides arenarius* (Kalyptorhynchia). Strong protective cushioning with branched braces inside. D. Section. E and F. *Bothriomolus balticus* (Otoplanidae, Seriata) from high energy beaches. Cement glands produce a cap which is pulled over the egg capsule like a slouch hat and whose rims adhere between grains of sand. (Giesa, Hoxhold, in Ax 1969)

This is taken a step further in *Diascorhynchides arenarius*. A 40–70-μm-thick cushion protects the egg from pressure and friction in the sandy bottom. A delicate covering is found in the periphery. The cushion's interior is scattered with branched, elastic braces (Fig. 76B–D).

The free-living Plathelminthes strive to anchor their eggs in the biotope. This is usually done with the help of special cement glands which surround the female genital pore. Very often, shell glands in combination with cement glands produce elastic egg stalks and glue them to the substrate with a broad basal plate (Fig. 76A). Evolution brought about exciting adaptations even at this late stage. In *Bothriomolus balticus* (Proseriata), an inhabitant of high energy beaches, extensive complexes of cement glands produce a large slouch hat. It is firmly connected with the egg stalk and pulled over the egg capsule as an additional protective hood (Fig. 76E,F).

Let us now justify the monophyly of the taxa Lecithoepitheliata and Prolecithophora.

Lecithoepitheliata

■ **Autapomorphy**

– Female gonads: epithelial-like arrangement of vitellocytes around one germocyte.

Based on the structure of the gonads already discussed, the marine Gnosonesimida (*Gnosonesima*) and the limnetic Prorhynchida (*Prorhynchus, Geocentrophora*) are placed together in the taxon Lecithoepitheliata.

We have interpreted the production of germocytes and vitellocytes in a mixed gonad to be a plesiomorphy. The covering of the single, young germ cells with a layer of yolk cells is, however, unique within the Neoophora (Fig. 77A). This special construction of the germovitellarium can thus be hypothesized to be a derived feature characteristic of the Lecithoepitheliata.

Prolecithophora

■ **Autapomorphy**

– Aciliated sperm with an extensive intracellular membrane system.

The taxon Prolecithophora includes numerous species, most of which are marine dwellers; *Plagiostomum lemani* is a well-known freshwater species.

A pharynx plicatus and a mixed gonad are plesiomorphies in the ground pattern of the Prolecithophora. It is not possible to deal with the diverse modifications of both organs in further detail here. The intracellular membrane system in the aciliated sperm is an impressive autapomorphy (Fig. 77C,D). It has been found in all species of the Prolecithophora which were electron-microscopically examined, and has not been discovered in any species which do not belong to the taxon. A detailed study of the intraspermial membrane system reveals that it has very different structures; it can, for example, take the shape of alveoli with stacks of membranes (*Archimonotresis*), of an even-folded band (*Pseudostomum*) or, in other cases, of fingerprints (*Multipeniata*) (EHLERS 1981, 1988; SCHMIDT-RHAESA 1993).

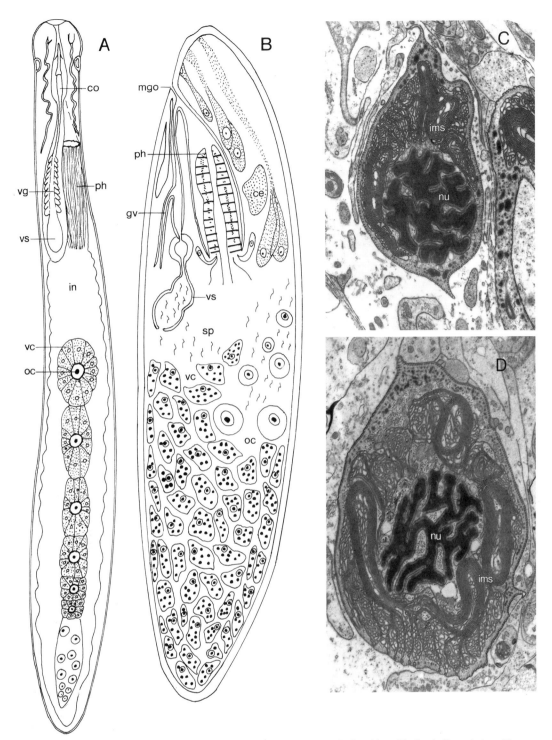

Fig. 77. Lecithoepitheliata and Prolecithophora. A. *Prorhynchus stagnalis* (Lecithoepitheliata). Dorsal view. The germocytes which grow from the back are each surrounded by a layer of vitellocytes. B. *Archimonotresis limophila* (Prolecithophora). Sagittal section. Male-female gonads with sperm and germocytes in the anterior area and following these in a caudal direction are vitellocytes. C-D. *Multipeniata* sp. (Prolecithophora). Ultrastructure of aciliated sperm. Extensive intraspermal membrane system in the form of fingerprints. ce = brain. co = copulatory organ. gv = germovitelloduct. ims = intraspermal membrane system. in = intestine. mgo = mouth-genital opening. nu = nucleus. oc = germocyte (oocyte). ph = pharynx. sp = sperm. vc = vitellocyte. vg = vesicula granulorum. vs = vesicula seminalis. (A Luther 1960; B Karling 1940; C and D Schmidt-Rhaesa 1993)

Seriata – Rhabdocoela

Since the kinship of the Lecithoepitheliata and Prolecithophora has not yet been clarified, the Seriata and the Rhabdocoela can, for the time being, be considered to be the highest ranking adelphotaxa of the Neoophora (Fig.78).

The Seriata have retained the original pharynx plicatus, but exhibit apomorphous traits, such as the follicular division of testes and vitellaria and their serial arrangement.

In contrast, the barrel-shaped pharynx bulbosus is a characteristic apomorphy of the Rhabdocoela.

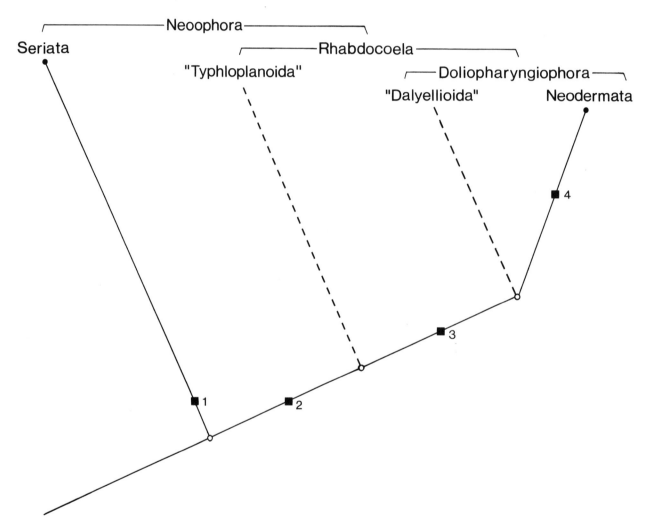

Fig. 78. Diagram of phylogenetic kinship within the plathelminth taxon Neoophora with the Seriata and the Rhabdocoela as sister groups. The diagram leads up to the parasitic Plathelminthes unified under the name Neodermata.

Seriata

■ **Autapomorphies**
(Fig. 78 → 1)

– Pharynx tubiformis.
The level of organization of the pharynx plicatus is plesiomorphous, whereas the stretching of the pharynx to a tube along the longitudinal axis of the body with the pharyngeal opening oriented towards the back is apomorphous (Fig. 80A,B).

– Division of the testes and vitellaria into numerous follicles. Arrangement of the follicles in even, longitudinal rows within the body.

The **Proseriata** (Fig. 79A,B) are more plesiomorphous than the Tricladida as concerns their intestine. They are represented by numerous species found in the benthic zone, where they unfold in the interstitial of marine sand. Thus, all over the world the Otoplanidae (*Otoplana, Bothriomolus, Parotoplana*) are the dominant animal group in high energy beaches.
Most of them are millimeter-sized with a rounded body. Some thread-like species from the taxa *Nematoplana* and *Coelogynopora* reach a length of about 1 cm.
In all species of the Proseriata which have been examined with the electron microscope, the lamellate rhabdites, which are part of the ground pattern of the Rhabditophora, are absent (SOPOTT-EHLERS 1985). This must be interpreted as a secondary absence, and can thus be considered as an autapomorphy of the Proseriata.

The **Tricladida** (Fig. 79C,D) are the best known Seriata – flat, centimeter-sized worms represented by marine species (*Procerodes, Uteriporus*), limnetic planarians (*Planaria, Dugesia, Dendrocoelum*), and terrestrial representatives (*Geoplana, Bipalium*) in the tropics.
The most noticeable autapomorphy is the intestine with three branches. One branch runs towards the front and two large diverticula are directed caudally. This gastrovascular system evolved as the result of an increase in size and volume which occurred within the Seriata. A completely different and very impressive characteristic feature occurs in ontogenesis. The germ of the Tricladida develops a temporary embryonic pharynx to absorb the extraembryonic yolk cells (Fig. 80C). Once this function has been fulfilled, the embryonic pharynx degenerates; it is in no way related to the pharynx plicatus of the adult.

Rhabdocoela

■ **Autapomorphy**
(Fig. 78 → 2)

– Pharynx bulbosus.
The maximum differentiation of the pharynx compositus (p. 162) occurs in the bulbous pharynx of the Rhabdocoela. Two evolutionary novelties in the structure of the pharynx bulbosus should be emphasized (Figs. 81,82):
1. A distinct septum completely separates the pharynx from the rest of the body.
2. The pharyngeal pocket is greatly shortened at the proximal end and covers only about a third of the pharynx.

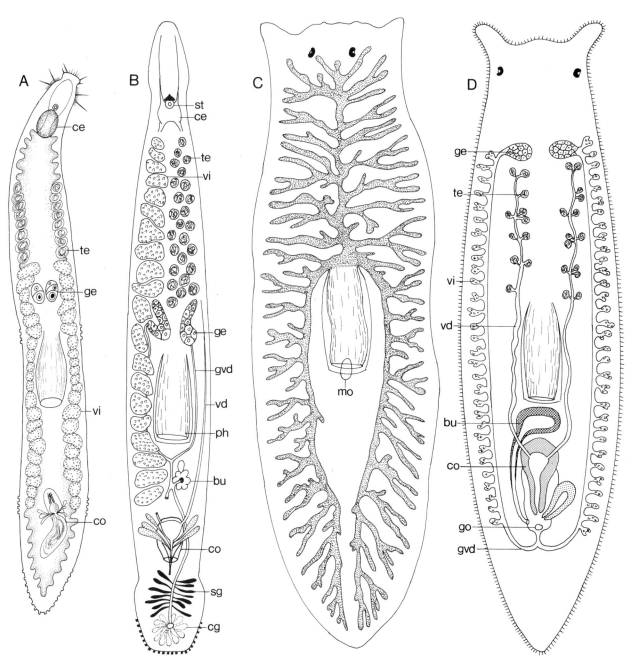

Fig. 79. Seriata. A. *Orthoplana kohni* as representative of the Otoplanidae (Proseriata). B. *Monocelis fusca* (Monocelididae, Proseriata). C. *Dendrocoelum lacteum* (Tricladida). Unpigmented freshwater species in which the construction of the intestine with three large branches is prominent. D. Diagram of the genital system of the limnetic Tricladida with follicular testes and vitellaria. bu = bursa. ce = brain. cg = cement glands. co = copulatory organ. ge = germarium. go = genital opening. gvd = germovitelloduct. mo = mouth. ph = pharynx. sg = shell glands. st = statocyst. te = testis follicle. vd = vas deferens. vi = vitellarium follicle. (A Ax and Ax 1967; B Giesa 1966; C de Beauchamp 1961b; D Bresslau 1928-33)

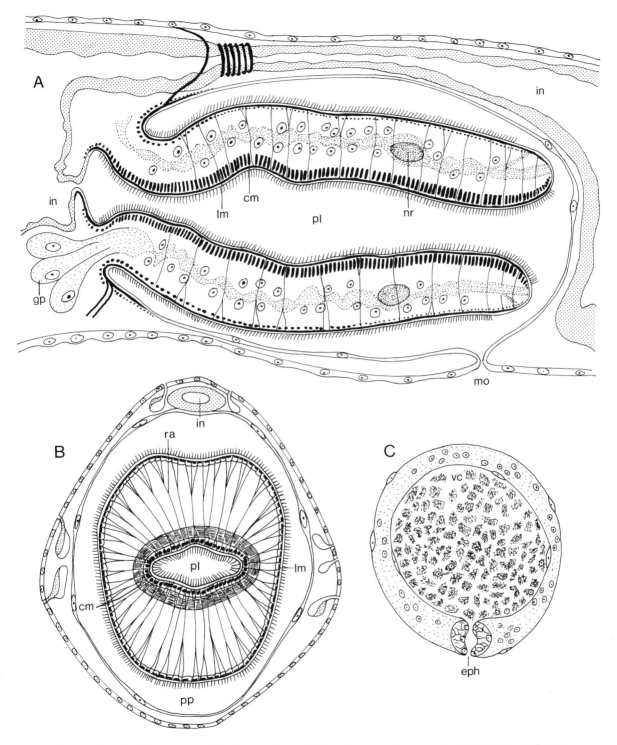

Fig. 80. Seriata. A and B. Construction of the pharynx plicatus tubiformis as an autapomorphy of the Seriata. A. *Otoplan-ella schulzi* (Proseriata). Sagittal section of the pharynx which lies horizontally in the body and is oriented towards the back. B. *Notocaryoplana arctica* (Proseriata). Cross section of the body at the height of the pharynx. C. *Dugesia gono-cephala* (Tricladida). Embryo with embryonic pharynx for the absorption of yolk cells. cm = circular muscle. eph = embryo-nic pharynx. gp = pharyngeal glands. in = intestine. lm = longitudinal muscle. nr = nerve ring. pl = pharyngeal lumen. pp = pharyngeal pocket. ra = radial muscle. vc = dissolved vitellocytes in the embryo. (A, B Ax 1956a; C Bresslau 1928-33)

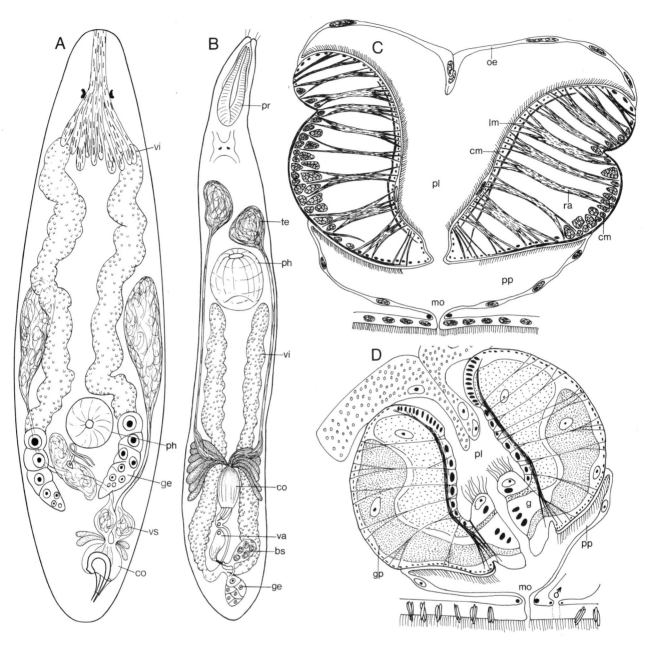

Fig. 81. "Typhloplanoida". A. *Proxenetes ampullatus*. Pharynx rosulatus in the second half of the body. B. *Proschizorhynchus gullmarensis* (Kalyptorhynchia, Schizorhynchia) with a proboscis composed of two muscle pincers at the anterior end. C and D. Construction of the pharynx rosulatus as plesiomorphous state of the pharynx bulbosus of the Rhabdocoela. In comparison with the pharynx plicatus, the following two characteristics should be pointed out: the complete separation against the body and the shortening of the pharyngeal pocket. C. Longitudinal section of the pharynx rosulatus of *Ciliopharyngiella intermedia*. The complete ciliation of the inner side against the pharyngeal lumen is original. D. Strongly differentiated pharynx rosulatus of *Maehrenthalia delamarei* with a grasping apparatus which is directed into the lumen. bs = bursa seminalis. cm = circular muscle. co = copulatory organ. g = grasping apparatus. ge = germarium. gp = pharyngeal gland. lm = longitudinal muscle. mo = mouth. oe = esophagus. ph = pharynx. pl = pharyngeal lumen. pp = pharyngeal pocket. pr = proboscis. ra = radial muscle. te = testis. va = vagina. vi = vitellarium. vs = vesicula seminalis. (Ax 1952, 1956c, 1966a, 1971)

"Typhloplanoida" – Doliopharyngiophora

We postulate that the just named features of the pharynx bulbosus evolved in a stem lineage common to all Rhabdocoela. Two widespread manifestations of the bulbous pharynx stand out within the Rhabdocoela – the pharynx rosulatus in the "Typhloplanoida", and the pharynx doliiformis of the Doliopharyngiophora. It is worth assessing these for the purposes of phylogenetic systematics.

"Typhloplanoida"

– Paraphyletic collection of the Rhabdocoela.
  No autapomorphies are known.

The pharynx rosulatus is located perpendicular to the ventral side. This is the original position of the pharynx compositus which was adopted into the stem lineage of the Rhabdocoela from the ground pattern of the Trepaxonemata. Contrary to conventional classification, the pharynx rosulatus, as a plesiomorphous feature, does not justify the taxon "Typhloplanoida".
Two examples may document steps in the evolution of the pharynx rosulatus. *Ciliopharyngiella intermedia* provides a very original state. It is only in this species that the pharynx is ciliated internally and externally. *Maehrenthalia delamarei* exhibits a highly modified pharynx rosulatus. Internal ciliation is considerably reduced here and at the distal end a powerful grasping apparatus has been developed for catching prey (Fig. 81C,D).

The **Kalyptorhynchia** form a monophyletic subtaxon of the "Typhloplanoida", characterized by the evolution of an anterior proboscis that can be pushed out terminally to catch prey. The original form of the proboscis as a uniform muscle bulb is realized in the euryhaline species *Gyratrix hermaphroditus*. The evolutionary unfolding of the Kalyptorhynchia took place in the sandy bottom of the ocean. Here the proboscis evolved to form a pincer made of two muscle tongues in the stem lineage of the Schizorhynchia (Fig. 81B), which later evolved additional hard hooks for grasping.

Doliopharyngiophora

■ **Autapomorphies**
  (Fig. 78 → 3)

– Pharynx doliiformis (Fig. 82)
  Three evolutionary novelties differentiate the pharynx doliiformis from the plesiomorphous state of the pharynx rosulatus:
  1. Shifting of the pharynx to the anterior end with either a terminal orifice, or an oral pore which slants slightly downward.
  2. Complete reduction in ciliation. The pharyngeal epithelium is aciliated both internally and externally.
  3. Shifting of the nucleated parts of the inner pharyngeal epithelium towards the interior; here they form an appendage at the juncture between pharynx and gut. This sub-feature is not realized consistently, and so it can be included only with reservations in the list of autapomorphies.

- Lack of the duo-gland adhesive organ.

The adhesive organ comprising three specialized cells, which we included in the ground pattern of the Rhabditophora, is absent in the Doliopharyngiophora including the free-living species. Since it is, however, present in the "Typhloplanoida", this absence must be a secondary state which could be explained as follows: the stem species of the Doliopharyngiophora underwent a change in biotope to a softer sediment, and so the adhesive organ became superfluous.

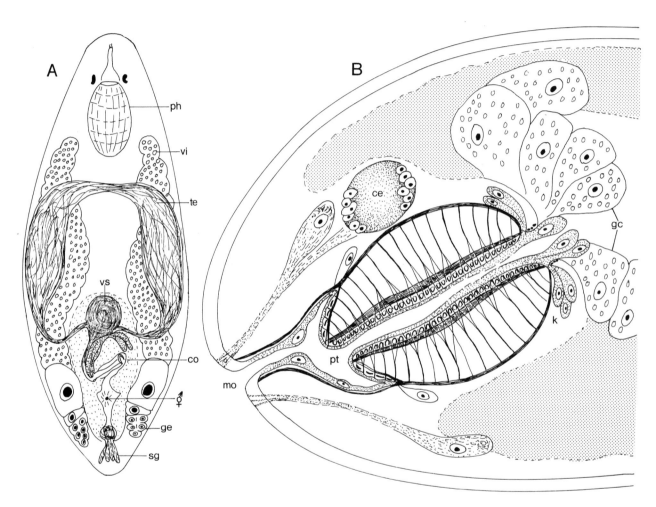

Fig. 82. Doliopharyngiophora. *Baicalellia subsalina* from the paraphyletic taxon "Dalyellioida". A. Organization of the 0.5-mm-long animal with pharynx doliiformis at the anterior end. B. Sagittal section of the pharynx doliiformis as an apomorphous state of the pharynx bulbosus of the Rhabdocoela. With subterminal mouth at the anterior end, short pharyngeal pocket, and complete lack of cilia. The nuclei of the inner pharyngeal epithelium form an appendage at the entrance to the intestine. ce = brain. co = copulatory organ. gc = glands of the intestinal epithelium. ge = germarium. k = appendage (nucleated cell bodies of the inner pharyngeal epithelium). mo = mouth. ph = pharynx. pt = pharyngeal pocket. sg = shell glands. te = testis. vi = vitellarium. vs = vesicula seminalis. (Ax 1954)

178

"Dalyellioida" – Neodermata

The free-living "Dalyellioida" and the classical parasite taxa are unified in the taxon Doliopharyngiophora. Whereas no autapomorphy has been found among the former, the parasites united under the Neodermata can clearly be shown to be elements of a monophylum.

"Dalyellioida"

– A taxon of conventional classification with predominantly free-living marine species (*Provortex, Baicalellia*) and freshwater representatives (*Dalyellia*). At present, they seem to be a paraphyletic collection of Doliopharyngiophora.

Neodermata

■ **Autapomorphies**
(Fig. 78 → 4)

– Degeneration of the ciliated epidermis.
The original multiciliated ectoderm of the free-living Plathelminthes is limited to larval stages in the taxon Neodermata: Trematoda – Cotylocidium and Miracidium; Monogenea – Oncomiracidium; Cestoda – Lycophora and Coracidium. The ciliated epidermis degenerates once the larval phase is completed.

– Neodermis.
Replacement of the larval ectoderm with a new, syncytial body covering (Fig. 83). The neodermis is formed by stem cells located inside the body. Cell processes of the neoblasts permeate the basal lamina and spread out beneath the ciliated epidermal cells. The latter degenerate and are sloughed off. The processes of the stem cells fuse together to form an aciliated syncytium. The nucleated cell bodies remain beneath the basal lamina. They are linked with the surface syncytium with cytoplasmic threads.

– Protonephridia with a two-cell weir.
The carrying structure for the ultrafilter of the ECM in the Trematoda, Monogenea, and Cestoda is composed of two rows of longitudinal rods. They are formed by the terminal cells and the first canal cell of the protonephridium. The rods of the canal cells push themselves from the outside over the rods of the terminal cell. This results in a complex double wreath consisting of longitudinal rods from two cells arranged alternately (Fig. 84B).
In the ground pattern of the Plathelminthes, on the other hand, the terminal cell alone forms the carrying structure of the filter. Among the free-living Rhabdocoela, the terminal cell is elongated distally to form a tube which is perforated all around. In a cross section view, this resembles a simple basket composed of a wreath of rods which surrounds the centrally positioned bundle of cilia of the terminal cell (Fig. 84A).
These clear facts indicate that the double basket is a further autapomorphy of the Neodermata.

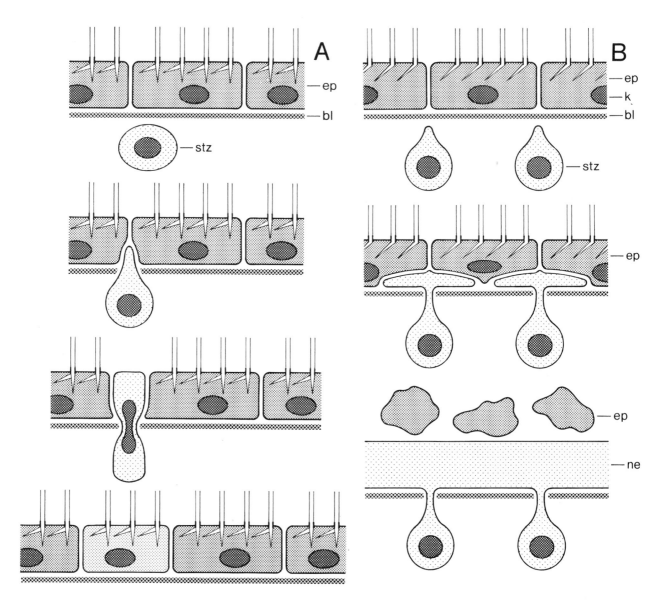

Fig. 83. Formation of a new ciliated cell in the primary epidermis of free-living Rhabdocoela (A) compared with the origination of a secondary neodermis in the parasitic Neodermata (B). Development proceeds from top to bottom. A. Stem cell (neoblast) from the inside of the body breaks out through the basal lamina and pushes itself between the differentiated epidermal cells. Here, it forms cilia and integrates itself into the epidermis as a new skin cell. B. Stem cells penetrate the basal lamina of the ciliated epidermis of larvae of the Neodermata (Trematoda, Monogenea, Cestoda) and spread themselves out below the skin cells or also between them. The ciliated epidermis then dies and is shed. The cell bodies of the neoblasts fuse together, building a syncytium, and form the neodermis; parts of the cells with the nuclei remain inside the body. The diagram also demonstrates the difference in the number of ciliary rootlets. A. There are two ciliary rootlets in the ground pattern of the Plathelminthes. B. Only the rostral rootlet is developed in the cilia of the larval epidermis of the Neodermata. bl = basal lamina. ep = epidermis. ne = neodermis. k = nucleus. stz = stem cell (neoblast). (Ax et al. 1989)

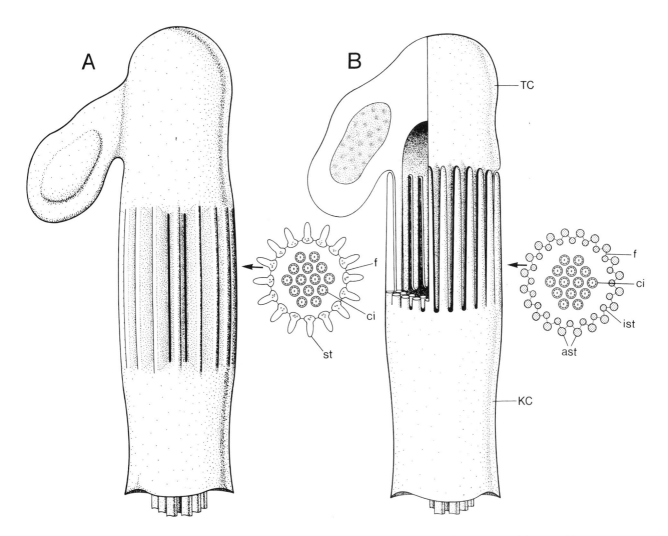

Fig. 84. Comparison of the terminal region of the protonephridia of free-living Rhabdocoela (A) with that of the parasitic Neodermata (B). Side view and cross section. A. In the ground pattern of the Rhabdocoela, the filter region of the proto-nephridium is produced of the terminal cell alone. The wall of the tube-shaped distal part of the cell is divided into numer-ous rods; between them are long slits. The filter matrix is spread out between the rods. There is a bundle of cilia inside the tube. B. In the Neodermata, the filter region is a compound product of the terminal cell (TC) and the first canal cell (KC). The canal cell pushes a second set of rods over the rods of the terminal cell forming together a double ring of rods with alternating arrangement. The filter matrix is positioned again between the rods. ast = external rods of the first canal cell. ci = cilium. f = filter matrix. ist = inner rods of the terminal cell in the Neodermata. st = rods of the terminal cell in free-living Rhabdocoela. (Ax et al. 1989)

Of all the apomorphous features of the taxon Neodermata, the unique degeneration of the plesio-morphous, ciliated epidermis and its replacement by a new, secondary body covering is specially worthy of mention. Since this feature is manifested among the Trematoda, Monogenea, and Cestoda, it can indisputably be postulated that the Neodermis evolved once in a stem lineage common to all of them.

The justification presented for the monophyly of a unity Neodermata compellingly implies the following: the parasitism of the Trematoda, Monogenea, and Cestoda developed once from a species of free-living Plathelminthes with a pharynx doliiformis. The evolution of the neodermis is obviously related to the evolution of this stem species of the Neodermata to a parasite within an invertebrate host organism.

Trematoda – Cercomeromorpha

In the trematode taxa Aspidobothrea and Digenea, the skin of the larvae is composed of ciliated ectodermal cells and parts of the neodermis. This is certainly an apomorphy as against the purely ectodermal skin in the larvae of the cercomeromorph taxa Monogenea and Cestoda. On the other

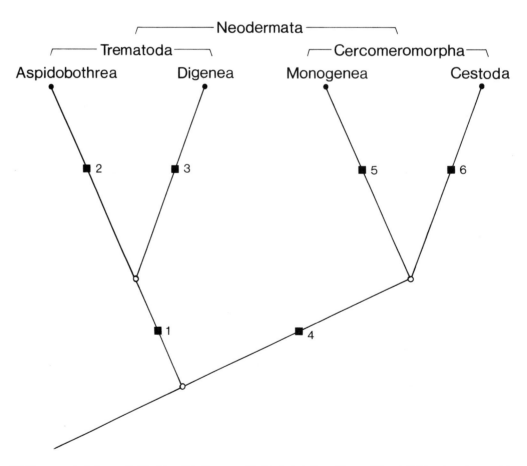

Fig. 85. Diagram of phylogenetic kinship within the parasitic Plathelminthes. The traditional division into the two "classes" of Trematoda and Cestoda is wrong. The Monogenea, which were incorporated within the Trematoda, are rather the sister group of the Cestoda, and are unified with them in the taxon Cercomeromorpha. The Trematoda and Cercomeromorpha are the highest-ranking adelphotaxa of the Neodermata.

hand, both these unities are characterized by sickle-shaped hooks at the posterior end of the larvae, which is an apomorphy compared with the lack of corresponding structures in the Trematoda (Fig.85).

Trematoda

■ **Autapomorphies**
(Fig. 85 → 1)

- The skin of the ciliated larvae is made up of ciliated ectodermal cells and portions of the neodermis.

- Primary host is a member of the Mollusca.

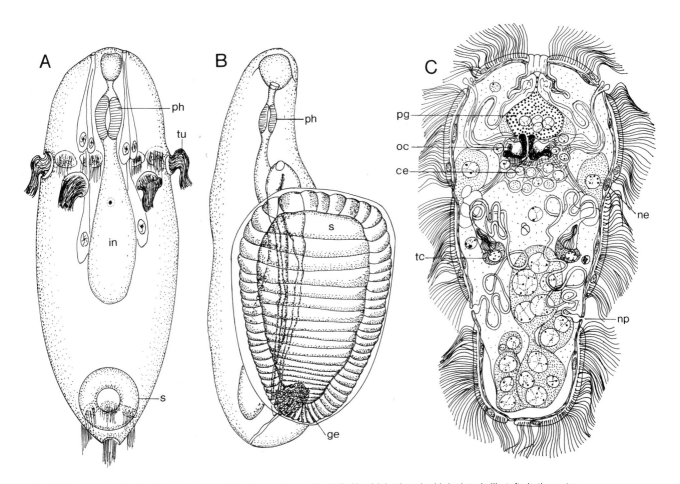

Fig. 86. Trematoda. A. Cotylocidium larva of *Cotylogaster occidentalis* (Aspidobothrea) with isolated cilia tufts in the anterior section of the body and at the posterior end; ventral sucker at the posterior end. B. Subadult individual of *Cotylogaster occidentalis* with well-developed ventral sucker. C. Miracidium larva of *Isthmiophora melis* (Digenea). Cross rows of ciliated epidermis are separated by strips of subepidermal neodermis. ce = cerebral ganglion. ge = germarium. i= intestine. ne = neodermis. np = nephroporus. oc = ocellus. pg = penetration gland. ph = pharynx. s = sucker (adhesive disc). tc = terminal cell of the protonephridium. tu = bundle of cilia. (A, B Fried and Haseeb 1991; C Dönges 1988)

In the miracidium of the Digenea there are distinct gaps between the ciliated cells arranged in transverse rows. Cytoplasmic processes of the subepithelial neodermis protrude into these gaps (Fig. 86C).

In the cotylocidium of the Aspidobothrea, ciliation is either greatly reduced (*Cotylogaster*) or completely absent (*Aspidogaster*). In *Cotylogaster occidentalis*, isolated bundles of cilia form a wreath around the anterior part of the body, and a second wreath just before the tail end; each bunch of cilia stems from a single epidermal cell (Fig. 86A). As for the rest, the body of the cotylocidium larva is covered by the neodermis.

The first host of the Digenea which is parasitized by the miracidium is almost always a mollusk (Gastropoda, Bivalvia). The primary host of the Aspidobothrea is also from the Mollusca, and it often remains the only host organism. One can conclude from this circumstance that a representative of the Mollusca was adopted as the phylogenetic primary host in the stem lineage of the Trematoda.

After the emergence of the vertebrates, the Digenea conquered the Gnathostomata as new hosts; they were obligatorily integrated into the life cycle of the parasites. A sequence of two hosts, the intermediate and definitive host, developed. The alternation of generations of the Digenea with three generations of different individuals is obviously related to this.

Aspidobothrea – Digenea

The most striking feature of the Aspidobothrea is a very large adhesive organ composed of numerous suction pits which occupies most of the ventral side. In contrast to this, the autapomorphies of the Digenea are the two-host cycle and the alternation of generations mentioned in the previous passage.

Aspidobothrea

■ **Autapomorphies**
(Fig. 85 → 2)

– Cotylocidium larva with an extensive reduction of epidermal ciliation. Ciliated cells are scattered primarily over the body; a secondary complete absence is possible.

– Cotylocidium larva with a round sucker positioned ventrally at the posterior end.

– Adult with adhesive organ on the ventral side which develops from the larval sucker and is greatly differentiated by ribbed and ring shaped elevations (Fig. 86B).

– Neodermis with microtubercles.
Microvilli are present in the Aspidobothrea as hemispherical microtubercles; they are not found in this form in any other taxon of the Neodermata.

The Aspidobothrea have an original, simple life cycle consisting of the cotylocidium larva and the adult organism. Primary hosts are marine and limnetic mollusks (snails, mussels) in which the animals mature. Single species have also been found in vertebrates. These species probably entered the vertebrates through mollusks in preadult stages or as adults.

Digenea

■ **Autapomorphies**
(Fig. 85 → 3)

– Two-host cycle with a mollusk as the intermediate host and a vertebrate as the definitive host.

– "Alternation of generations" with three generations of different individuals.
1. Generation – miracidium + sporocyst
Miracidium as a free-swimming ciliated larva and sporocyst with vegetative reproduction in a mollusk.
2. Generation – redia
Redia with vegetative reproduction in a mollusk.
3. Generation – cercaria + hermaphrodite
Cercaria as a free-swimming larva with a tail and hermaphrodite with bisexual reproduction in a vertebrate.

– Ectodermal cells of the miracidium arranged in transverse rows.
The ciliated cells are separated by narrow strips of cytoplasm of the subepidermal neodermis resulting in an evenly striated formation (Fig. 86C).

The main autapomorphies of the Digenea – the two-host cycle and the alternation of generations – are well illustrated in the large liver fluke *Fasciola hepatica* (Fig. 87). The problematical use of the terms individual, life cycle and generation will be discussed later (p. 188).

The hermaphroditic trematode *Fasciola hepatica* lives in the bile ducts of the liver of various mammals, especially in ruminants (cattle). Let us start the life cycle with the fertilized egg, which is laid by the hermaphrodite. The egg leaves the vertebrate host through the intestine. In order for the miracidium to hatch, it must reach an aquatic environment. The lung snail *Lymnaea (Galba) truncatula* is a common intermediate host. After shedding its ciliated epidermis, the miracidium enters the snail and grows into a sporocyst. This is a simple sac without an intestine which absorbs nutrients through the neodermis of the body surface. The first generation ends with the formation of the sporocyst.
Several redia develop from neoblasts (stem cells) of the sporocyst. The single redia now has a pharynx and a rod-shaped intestine. The redia form the second generation. In the case of *Fasciola hepatica*, vegetative reproduction results in a variable number of further "generations" of redia. They are not part of the basic pattern of the Digenea, however, and can be ignored here if only individuals with different structures are interpreted as representatives of different generations.

The individual redia produces cercariae which are the larvae of the third generation. This is once again by asexual reproduction through stem cells. The presence of the oral and ventral suckers and a bifurcated intestine in the cercaria already indicate essential features of the hermaphroditic trematode. A characteristic larval feature is the tail. The cercaria leaves the water snail and represents the second free-living phase in the life cycle of the large liver fluke after the miracidium. Whereas the cilia were used for locomotion in the former phase, the tail is used in the latter phase. The cercaria comes to rest on plants, encysts itself, and sheds its tail. The encapsulated

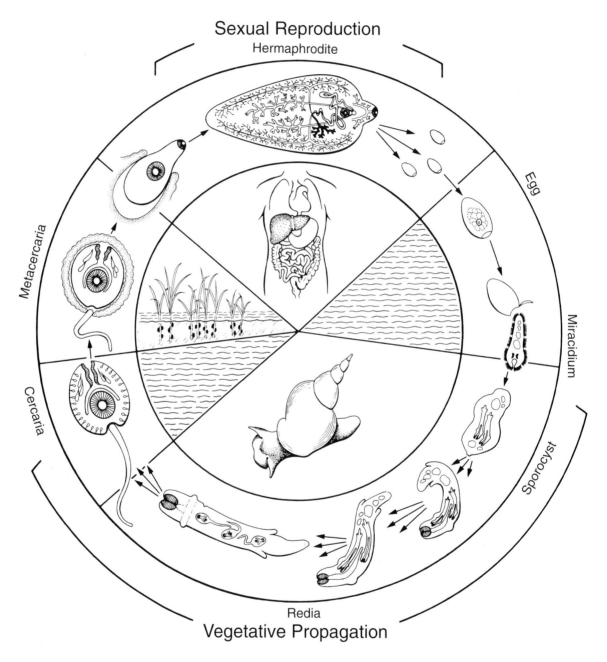

Fig. 87. "Alternation of generations" of the Digenea. Life cycle of the large liver fluke *Fasciola hepatica* with the succession of three generations of different individuals. 1. Generation = miracidium + sporocyst. 2. Generation = redia. 3. Generation = cercaria + hermaphrodite. (Ehlers 1985). It is still unclear whether the sporocyst and the redia reproduce purely vegetatively from neoblasts (stem cells) or if parthenogenesis with meiosis occurs (Dönges 1988).

metacercaria is ingested with the plant by the definitive host. The metacercaria slips into the intestine, bores through the intestinal wall into the body cavity, and penetrates the liver from the outside. The metacercaria develops into a mature hermaphrodite here. This ends the third generation. Fertilized eggs from bisexual reproduction begin the next life cycle.

The central "motif" of the Digenea – a life cycle consisting of individuals with different methods of reproduction, and the parasitism of two hosts – was changed by evolution, resulting in a fascinating number of variations. Two examples will illustrate the amazing ways of adaptation that have occurred.

The small liver fluke *Dicrocoelium dentriticum* (Fig. 88A-C) completely freed itself from water, and became a truly terrestrial animal. The hermaphrodites live in the bile ducts of the liver of ruminants (sheep, deer). When the egg is laid on the ground, it contains a differentiated miracidium, which must wait until it is ingested by xerophilous terrestrial lung snails like *Zebrina detrita* or *Helicella* species. The miracidium hatches in the intestine of the snail, and reaches the interior of the body (digestive glands) through the intestinal wall, where it grows into a sporocyst (mother sporocyst). A second sporocyst (daughter sporocyst) develops instead of the redia. It produces cercaria which are incorporated in snail slime and excreted through the lungs. *Dicrocoelium dendriticum* requires a secondary intermediate host to continue its life cycle. Certain species of ants (taxon *Formica*) eat the balls of slime. The cercaria bore through the stomach wall while shedding their tails. Almost all of them migrate towards the posterior end and encyst themselves to become infectious metacercariae. One single cercaria performs a fantastic "social" achievement. It infects the ant's subesophageal ganglion, blocking its own further development. At the same time, however, it causes a remarkable behavioral change in the ant. In the evening, the ant clings to the tip of a plant, ceremoniously offering itself to be eaten by the ruminants. This leads us to the last characteristic feature. Whereas the metacercariae of *Fasciola hepatica* take the intricate route through the body cavity to bore their way into the liver, the metacercariae of *Dicrocoelium dentriticum* migrate from the intestine directly through the ductus choledochus into the bile ducts.

Three species of the taxon *Schistostoma* cause bilharziasis in humans in the tropics – *Schistosoma haematobium*, *S. mansoni*, and *S. japonicum*. Their life cycle is dictated by the conquest of a new territory in the vertebrate host, namely the circulatory system. Since encounters may be difficult here, the secondary gonochoric *Schistosoma* species have developed a remarkable measure to ensure an intimate, long-term connection of the sexes. The male folds the sides of its body ventrally inwards forming a groove with the attractive designation canalis gynaecophorus, in which he carries the thinner, more rounded female (Fig. 88D). How do the eggs get from the circulatory system to the outside? Again, nature has come up with a remarkable answer. During the production of the egg capsule in the ootype, the wall receives a special tooth or spur with which the endothelium of the blood vessels can be punctured – and this appropriately takes place around the urinary bladder or the intestine. The eggs, depending on the species, reach the outside along with urine or excrements. The miracidium needs water in order to hatch. There is nothing unusual about the further stages in the life cycle that occur in freshwater snails – formation of the mother sporocyst, daughter sporocyst, and cercaria. Finally, the fork-tailed cercaria offers a special contribution to the completion of the life cycle. It bores directly through the human skin into its definitive host and reaches the blood vessels through venous capillaries in the periphery of the body.

Individual – Life cycle – Generation

At this point I would like to present a few ideas on the use of the terms individual, life cycle, and generation in connection with the Digenea.

The freshwater mussel *Anodonta anatina* will be used as an example of the usual meaning. The

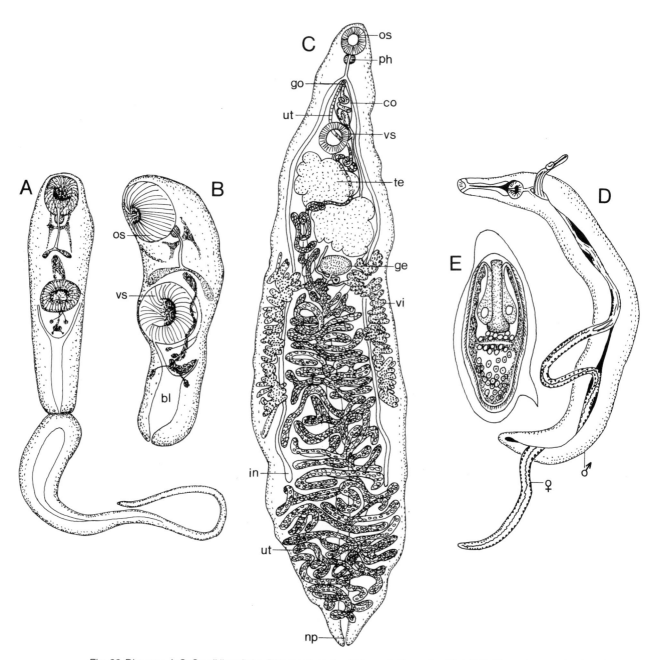

Fig. 88. Digenea. A-C. Small liver fluke *Dicrocoelium dendriticum*. A. Cercaria with tail. B. Young trematode taken from the definitive host. C. Adult with male and female genital organ. D. *Schistosoma haematobium*. Gonochoric. The male holds the female with the sides of its body. E. *Schistosoma mansoni*. Egg capsule with tooth; inside the miracidium. bl = urinary bladder. co = copulatory organ. ge = germarium. go = genital opening. in = intestine. np = nephroporus. os = oral sucker. ph = pharynx. te = testis. ut = uterus. vi = vitellarium. vs = ventral sucker. (A, B, E Piekarski 1954; C, D Hyman 1951)

glochidium larva hatches from the egg within the mother organism. It becomes a parasite in the skin of a freshwater fish, where it develops into a young mussel. It is released from the skin, lives on the floor of the water body as a filter feeder, and grows into an adult. This example describes a single individual of *Anodonta anatina* and one life cycle. However, the life cycle of the individual manifests itself in highly divergent forms, from the egg through the ectoparasitic glochidium to the adult; and this individual itself belongs to a generation of individuals set between the parent's generation and the generation of the prospective offspring. The individual is thus part of a generation of coexisting individuals in chronologically succeeding populations of the freshwater mussel *Anodonta anatina*.

How does this relate to the digenic Trematoda?

The entire sequence of miracidium–sporocyst–redia–cercaria–hermaphrodite can be shown to be one, connected life cycle. There is, however, no doubt that this is not the life cycle of one individual. If the sporocyst, redia, and hermaphrodite each reproduce, then they must be referred to as separate individuals. Hundreds of hermaphroditic trematodes which stem from a single miracidium simply cannot be seen as parts of one individual. In other words, a superindividual unity composed of several individuals represents a life cycle in the species of the Digenea, whereby the miracidium is obviously the larva of the sporocyst and the cercaria the larva of the hermaphrodite.

This brings us to the issue of what exactly a generation comprises. Does this superindividual unity of several individuals (miracidium + sporocyst, redia, cercaria + hermaphrodite) form a single generation, or is it composed of a succession of several generations which correspond to the number of individuals? This is where the obligatory change between different methods of reproduction plays a part in these considerations – the change between asexual reproduction (sporocyst, redia) and bisexual reproduction (hermaphrodite). If we take the definition of "alternation of generations" to be the "development of an organism by different methods of reproduction" (HENTSCHEL and WAGNER 1990), then the single superindividual unity of the Digenea is an "organism" comprising three generations. Is this a satisfying interpretation? Perhaps the diversity of living beings confronts us with circumstances which do not always allow entirely unambiguous terminology.

Cercomeromorpha

■ **Autapomorphy**
(Fig. 85 → 4)

– Sickle-shaped hooks (caudal hooks) at the posterior end of ciliated larvae. Number in the ground pattern = 16.

Monogenea – Cestoda

The larvae of the Monogenea (oncomiracadium) and the Cestoda (lycophora, coracidium, oncosphaera) all have sickle-shaped hooks on the posterior body pole. The primary number of 16 hooks appears only among the Monogenea (Fig. 89). Within the Cestoda, the lycophora larva of the Gyrocotylidea and Amphilinidea has 10 hooks. The number is reduced to six in the coracidium and the oncosphaera of the Cestoidea (Fig. 90).

Based on congruences in structure, position, and intracellular formation in oncoblasts, it can be postulated that the hooks evolved once in a stem lineage common to the Monogenea and Cestoda. The existence of hooks can be interpreted as a synapomorphy of the Monogenea and Cestoda. This means that the Monogenea, contrary to traditional classification, do not belong to the Trematoda, but form a monophylum together with the tapeworms.

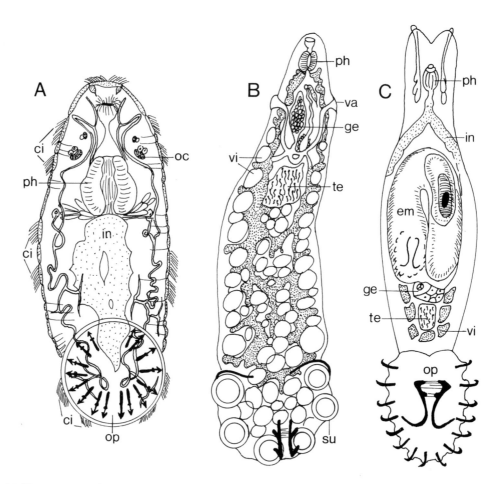

Fig. 89. Monogenea. A. Oncomiracidium larva of *Polystomum.* Epidermis with three complexes of ciliated cells. Caudal adhesive apparatus (opisthaptor) with 16 hooks. B. Adult of *Polystomum integerrimum.* Caudal adhesive organ (opisthaptor) with six suckers and one pair of hooks. C. *Gyrodactylus elegans.* Adult. Opisthaptor with peripheral ring of small hooks and two large, median hooks. ci = complexes of ciliated cells. em = embryo. ge = germarium. in = intestine. oc = ocellus. op = opisthaptor. ph = pharynx. su = opisthaptorial sucker. te = testis. va = vagina. vi = vitellarium. (A Hyman 1951; B,C Dönges 1988)

Monogenea

■ **Autapomorphies**
(Fig. 85 → 5)

– Oncomiracidium is not completely ciliated. Ciliated epidermal cells form three cilia complexes which succeed each other in the longitudinal axis of the body. An aciliated syncytial epidermis lies between them (Fig. 89A).
This mode of differentiation differs fundamentally from the partial ciliation of the miracidium and cotylocidium of the Trematoda. The aciliated areas of the oncomiracidium are not a neodermis, but form a primary ectodermal body covering in which the cilia are reduced and the cell walls have disappeared.

– Oncomiracidium has two pairs of rhabdomerous photoreceptors in pigment cups.

In comparison with the Digenea, the name Monogenea refers to the alternative in the life cycle. The Monogenea have a simple life cycle consisting of one larva (oncomiracidium), and the adult. The life cycle comprises one individual. This is the original state, which also forms part of the ground pattern of the Cercomeromorpha.

In contrast to the two-host cycle of the Digenea, the single species of the Monogenea have only one host. The host organism is primarily a gnathostome fish on which the Monogenea live as ectoparasites. *Dactylogyrus hastator* and *Gyrodactylus elegans* are two parasites which are dreaded by carp breeders. Several Monogenea penetrate into peripheral body cavities such as the urinary bladder or cloaca, where they evolve into endoparasites. While preparing frogs (*Rana*), one can come across *Polystomum integerrimum*. This parasite lives in the urinary bladder and is equipped with a powerful adhesive organ with six suckers and a pair of hooks.

Cestoda

■ **Autapomorphies**
(Fig. 85 → 6)

– Larvae and adults without an intestine.
The complete reduction of the gut is the outstanding characteristic feature of the tapeworms. Nutrition in the form of liquid substance from the host is absorbed through the neodermis.

– Larvae in the ground pattern have only ten hooks (Fig. 90A).

– Ciliated epidermis of the larvae is a syncytial tissue.

Three interesting taxa of "monozoic" tapeworms – Gyrocotylidea, Amphilinidea and Caryophyllidea – with a single hermaphroditic genital organ contrast to the Eucestoda, which has several sexual organs in one individual. The primary monozoic flatworms do not,

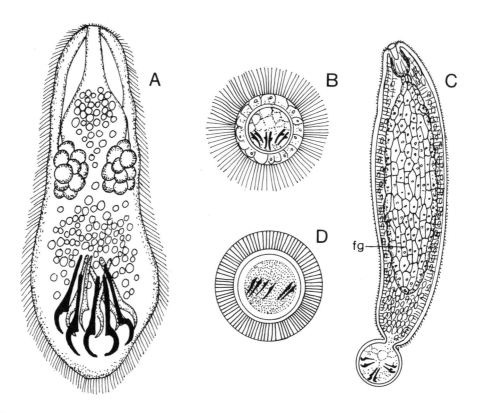

Fig. 90. Cestoda. Larvae. A. Lycophora larva of the Gyrocotylidea with ten hooks in the posterior end (feature of the ground pattern of the Cestoda). B. Oncosphaera in the form of the primary ciliated coracidium larva of *Diphyllobothrium latum*; six hooks form an autapomorphy of the Cestoidea. C. Procercoid of *Diphyllobothrium latum* with the six hooks of the Oncosphaera at the posterior end. D. Oncophaera in *Taenia saginata* with a radially striped covering. fg = frontal glands. (A Coil 1991; B-D Hyman 1951)

however, form a monophyletic unity, but are related to the Eucestoda in a sequence that can be precisely determined (Fig. 91).

The **Gyrocotylidea** are the sister group of all other tapeworms assembled under the name **Nephroposticophora**. Whereas in the Gyrocotylidea, the excretory pores of the paired proto-nephridia are located dorsally in the anterior of the body (plesiomorphy), the shifting of the opening of the protonephridial system to the posterior end is the essential autapomorphy of the Nephroposticophora.

Characteristic apomorphies of the Gyrocotylidea include a sucker-like structure at the anterior end, and the rosette-shaped adhesive organ at the posterior end (Fig. 91A). The lycophora is the original larval form (Fig. 90A). The Gyrocotylidea live as parasites in the intestine of Holocephali, a unity of the Chondrichthyes. *Gyrocotyle urna* lives in *Chimaera monstrosa*.

The **Amphilinidea** form the adelphotaxon of the **Cestoidea** (Caryophyllidea + Eucestoda). The larval stage in the former is once again the plesiomorphous lycophora larva with ten hooks. The reduction to six hooks in the larva of the Cestoidea is, on the other hand, an autapomorphy of this unity.

The extremely leaf-shaped body and a proboscis (apical organ) at the anterior end of the Amphilinidea are autapomorphies (Fig. 91B). The Amphilinidea have a two-host cycle. The lycophora

192

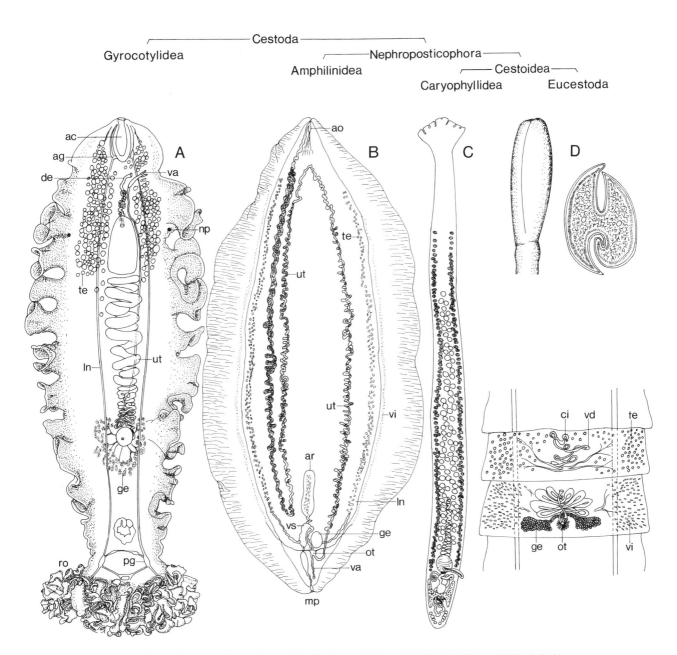

Fig. 91. Representatives of the highest-ranking subtaxa of the Cestoda. A. *Gyrocotyle fimbriata* (Gyrocotylidea). B. *Neso-lecithus africanus* (Amphilinidea). C. *Caryophyllaeus laticeps* (Caryophyllidea). D. *Diphyllobothrium latum* (Eucestoda). Side view and cross section of head with slit-shaped suckers. Below: proglottids with genital organs. Upper proglottid shows only the male system, lower shows only the female organs. ac = acetabulum (sucker). ag = frontal ganglion. ao = apical organ (proboscis). ar = accessory receptaculum seminis. ci = cirrus. de = ductus ejaculatorius. ge = germarium. ln = longitudinal nerve. mp = male genital pore. np = nephropore. ot = ootype. pg = posterior ganglion. ro = rosette-shaped adhesive organ. te = testis. ut = uterus. va = vagina. vd = vas deferens. vi = vitellarium follicle. vs = vesicula seminalis. (A Coil 1991; B Dönges 1988; C Joyeux and Baer 1961; D Piekarski 1954; Dönges 1988; phylogenetic kinship diagram at the top Ehlers 1985)

larvae parasitize Amphipoda, which serve as an intermediate host. The definitive hosts are the Acipenseroidea (sturgeon) and Teleostei (bony fish), where the adults live in the body cavity.

The **Caryophyllidea** as the third group of "monozoic" flatworms, and the **Eucestoda** form the last sister group pairs of the Cestoda we will discuss here.

Characteristic features of the Caryophyllidea (Fig. 91C) are the existence of secondary, monociliary sperm and the conquest of limnetic Annelida (in place of crustaceans) as the intermediary host. Freshwater bony fish are the definitive hosts; the intestine is the biotope.

The **Eucestoda** are the ultimate tapeworms. Their most prominent autapomorphy is the series of hermaphroditic genital organs in the longitudinal axis of one individual.

A phylogenetic systematization does not yet exist for the Eucestoda. Only a few allegedly original traits in the ground pattern of the unity can be discussed here.

Cyathocephalus truncatus or *Ligula intestinalis* are examples from a whole series of species which show no kind of external division. A strobila, which is customarily ascribed to the tapeworms as a chain of proglottids separated from each other by furrows, is obviously not part of the ground pattern of the Eucestoda; such a strobila evolved only within the unity.

A similar statement probably also applies to the "scolex with adhesive organs". *Spatheobothrium simplex* is a species of the Eucestoda with a round anterior end lacking adhesive organs which, in addition, is equipped with numerous sets of genital organs in an undivided body. The formation of a scolex among the Eucestoda obviously began with the evolution of two slit-shaped suctorial pits such as are seen in *Ligula intestinalis* or *Diphyllobothrium latum* (Fig. 91D).

Other original features of the Eucestoda are the existence of swimming larvae and the bond to an aquatic intermediate host. The ectoderm of the oncosphaera is primarily completely ciliated; this original form of larva is termed coracidium. When in the course of evolution, eggs were laid on land, the ciliated epithelium developed into a resistant covering which is radially striped in *Taenia saginata* and *Taenia solium* as a result of keratin rod deposits (Fig. 90B,D).

The life cycle of the Eucestoda can cover several larval stages while parasitizing different host organisms. The cycle is primarily the development of an individual without the alternation of generations. Asexual reproduction of the cysticerus in *Multiceps* and *Echinococcus* is a secondary phenomenon within the Eucestoda.

We will use the fish tapeworm *Diphyllobothrium latum* as an example of a life cycle with the original coracidium larva. The adult parasitizes humans and a wide spectrum of fish-eating mammals. The eggs which have been laid in the intestine must reach water before the coracidium (larva 1) can hatch. Here, they are eaten along with plankton by certain Copepoda of the taxa *Cyclops* or *Diaptomus*. The ciliated epithelium is digested in the intestine of the Copepoda. After its release, the oncosphera migrates into the body cavity of the first intermediate host where it grows into a proceroid (larva 2). On the posterior portion of the body, the elongated proceroid retains the six hooks of the oncosphera (Fig. 90C). In order for the life cycle to continue, the infected crustaceans must be eaten by fish (second intermediate host). The proceroid develops into a plerocercoid (larva 3), which now displays the suctorial pits of the adult. The plerocercoid penetrates the intestinal wall with the help of a secretion from a glandular complex in the anterior of the body and enters into the musculature of the fish, ready to be taken up by the definitive host. The transfer of the plerocercoid to predatory fish as an additional intermediate host is also possible.

Xenoturbellida

– A Eumetazoan Taxon with Unclarified Kinship –

■ **Autapomorphies**

– Epidermal cilia with two perpendicular ciliary rootlets (next to the caudal rootlet) oriented downwards, which probably arose from the splitting of the rostral rootlet.

– Narrow, ciliated epidermal cells each with a supporting fiber composed of bundles of filaments.

– Subepidermal membrane complex comprising an outer and inner basal lamina and a middle layer of striated filaments.

– Intraepidermal "statocyst" with freely moving, monociliated cells which swim about inside the organ.

Xenoturbella bocki, found on soft marine bottoms 40–100 m deep, is the only species of a high ranking taxon that was categorized under the Plathelminthes as a "suborder" by WESTBLAD (1949). It is a relatively large worm, measuring several centimeters in length. Anterior and posterior ends are tapered, and the body is slightly dorso-ventrally flattened. The organism is white, colorless.
Deep ciliary furrows are laterally in the completely ciliated epidermis; an additional furrow runs around the middle of the body (Fig. 92A,E).
Just before the middle of the body, **a simple, ventral oral pore** leads into the sack-shaped intestine. A pharynx is not differentiated. The intestine does not have an anus and the intestinal cells are aciliated.
The **nervous system** is an intraepithelial nerve plexus which lies completely in the base of the epidermis. *Xenoturbella bocki* has neither a brain nor concentrations in the form of nerve cords. A **statocyst** with 10–40 "statoliths" is embedded in the nerve plexus at the anterior end.
Beneath the epidermis and located inwardly from a subepidermal membrane complex made of ECM lies the body wall musculature with outer circular muscles and inner longitudinal muscles. The intestine is also surrounded by a network of muscles.
A weak, loose connective tissue (parenchyma) is located between the ectoderm and endoderm. This tissue with extensive extracellular spaces is braced by parenchyma muscles.

Xenoturbella bocki is a **hermaphrodite** with an extremely simple sexual organization. There are no localized gonads, no copulatory organs, nor is there a genital pore. Egg cells and sperm develop in the parenchyma around the intestine. The ripe, entolecithal eggs penetrate the intestinal epithelium and are released through the mouth. Sperm is also probably released in the same way, but has not as yet been observed.
The structure of the sperm is particularly interesting. *Xenoturbella bocki* has the original sperm of the Metazoa (p. 54) with a rounded head, a protruding middle piece with four mitochondria and a long cilium (Fig. 92G). The strict correlation between this sperm and the mode of external fertilization strongly leads to the conclusion that the gametes unite outside the body. The ontogenesis is unknown.

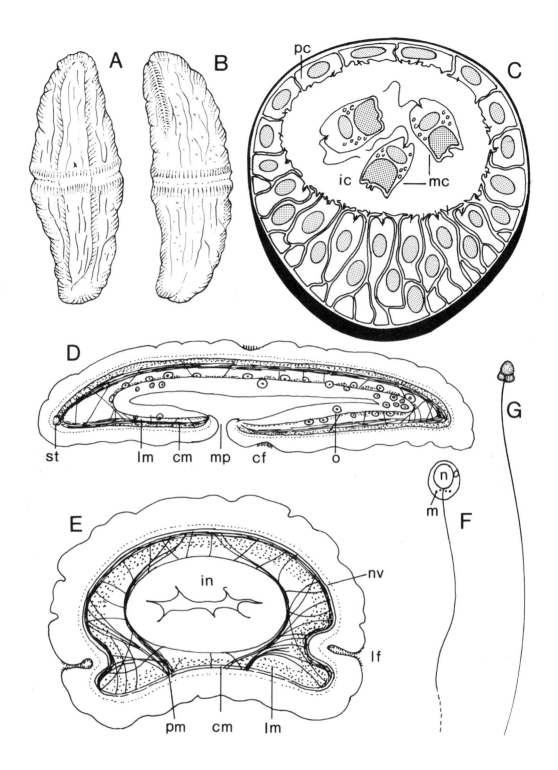

Fig. 92. Xenoturbellida. *Xenoturbella bocki.* A and B. Habitus of living organisms. C. Cross section of the "statocyst". Increase of parietal cells in the ventral region. Moveable, monociliary cells in the central cavity. D. Saggital section of the body. Mouth on the ventral side. E. Cross section of the body. F. Late spermatide. G. Mature sperm. cf = circular furrow. cm = circular muscle. ic = intracapsular cavity. in = intestine. lf = longitudinal furrow. lm = longitudinal muscle. m = mitochondrium. mc = moveable cells. mp = mouth pore. n = nucleus. nv = nerve plexus. o = oocyte. pc = parietal cell. pm = parenchyma muscles. st = "statocyst". (A, B, D, E, G Westblad 1949; C Ehlers 1991; F Franzén 1956)

196

Studies of the ultrastructure have revealed the structure of the epidermis and the statocyst (PEDERSEN & PEDERSEN 1986, 1988; FRANZEN and AFZELIUS 1987; EHLERS 1991).

The **epidermis** reaches a thickness of several hundred micrometers, which is quite unusual. It consists of highly multiciliated supporting cells, as well as different monociliated and aciliated glandular cells.

The multiciliated epidermal cells with a nucleus are extremely slender. They extend through the entire epidermis from the surface to the basal lamina. The free cilia exhibit the normal 9 x 2 + 2 microtubule pattern along most of its length. The tapered tip, however, shows a 5 x 2 + 2 pattern. The peripheral double microtubules 4–7 end approximately 0.5–1.5 µm before the tip, at which point the cilium is indented (Fig. 93B). In the distal shaft, the two central microtubules shift to the

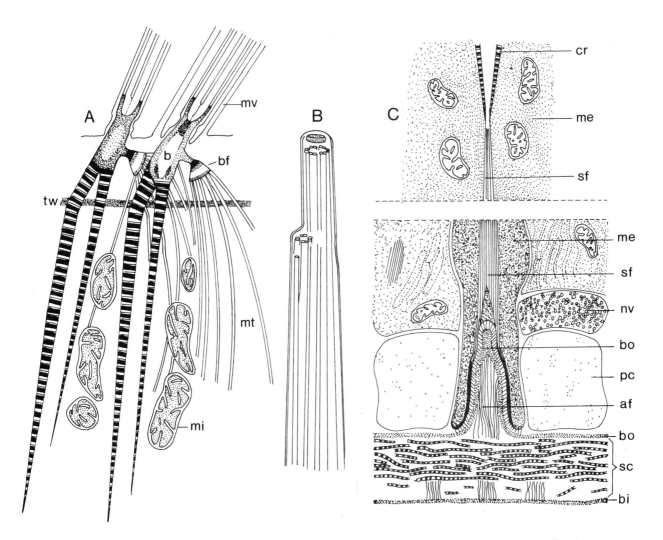

Fig. 93. Xenoturbellida. *Xenoturbella bocki.* Ultrastructure of cilia and epidermis. A. Sagittal section of two cilia. Basal body with two long rootlets which run perpendicularly into the cell, and a short rootlet which runs in a caudal direction. B. Tip of the cilium with shaft region. C. Diagram of selected features of the epidermis and of the subepidermal membrane complex. af = anchor filaments. b = basal body. bf = caudal ciliary rootlet (basal foot). bi = basal lamina (inner border). bo = basal lamina (outer border). cr = ciliary rootlet. me = multiciliary epidermis cell. mi = mitochondrium. mt = microtubules. mv = microvilli. nv = nerve process. pc = pillar cell. sc = subepidermal membrane complex. sf = supporting fibrils. tw = intracellular texture (terminal web). (A, B Franzén and Afzelius 1987; C Pedersen and Pedersen 1988)

periphery to form a wreath with the remaining double microtubules 1, 2, 3, 8 and 9. Two cross-striated ciliary rootlets per cilium run approximately perpendicular from the basal body into the epidermal cell. They are positioned very close together, and are longer than the free part of the cilia. An additional cross-striated ciliary rootlet ("basal foot") points backwards. It is short and widens distally to form an end-plate; 15–25 caudally oriented microtubules originate here (Fig. 93A).

Characteristic differentiations of the ciliated epidermal cells are intracellular supporting fibrils composed of bundles of filaments. A multiciliated supporting cell contains one single supporting fibril. The fibrils are concentrated in the lower two-thirds of the epidermis. They are attached to bell-shaped elevations of the outer basal lamina. The number of filaments is greatly reduced towards the outside; the few remaining filaments are in contact with the tips of ciliary rootlets (Fig. 93C).

An unusally thick "subepidermal membrane complex" of approximately 5 µm lies beneath the epidermis. A strong layer of striated filaments, probably composed of collagen, is found between an outer and inner basal lamina.

The **statocyst** of *Xenoturbella* (Fig. 92C) is peripherally surrounded by a strong capsule, which is completely lined with parietal cells that increase in number in the ventral area. The parietal cells are monociliated cells with a short cilium and an accessory centriole next to the basal body. Additional monociliated cells are found free and in greater numbers in the intracapsular cavity. They have a nucleus, a large vacuole which contains statolith-like concretions, and a long cilium. The cilium serves as a locomotory organelle, allowing the free cells in the statocyst to swim about randomly, rotating consistently around their longitudinal axis. After seeing this, it is hard to conceive of the statoliths functioning in a gravity sensory organ.

Phylogenetic Kinship

The idea that *Xenoturbella bocki* belongs to the Plathelminthes, and particularly that it is related with the Acoelomorpha, has been repeatedly rejected. It was, however, recently given renewed support (ROHDE et al. 1988; HASZPRUNAR, et al. 1991).

Let us examine the question, using the instruments of consistent phylogenetic systematic argumentation.

Hermaphroditism is an apomorphous correspondence with the Plathelminthomorpha. However, this alone does not tell us very much, since hermaphroditism evolved many times convergently within the Metazoa.

The ultrastructure of the epidermal cilia reveals one apomorphous congruence with the Acoelomorpha. This is the identical shaft region of the cilia tip, where the double microtubules 4-7 are lacking. However, the short caudal ciliary rootlet of *Xenoturbella* and the Acoelomorpha can be compared only to a limited extent. In *Xenoturbella*, it ends in a wide end-plate, whereby here the division of the rootlet into two lateral fiber bundles, characteristic of the Acoelomorpha, is absent. Finally, the construction of the rostral rootlet is completely different in this complex of features. The evolution of two lateral rootlets from the angled section of the rostral rootlet in the Acoela is in direct contrast to the division of the rostral rootlet into two elements positioned behind each other in *Xenoturbella*.

Moreover, the structure of the ciliated epidermal cells with long supporting fibrils is a feature peculiar to *Xenoturbella*, and unparalleled in the Plathelminthes. The highly evolved subepidermal membrane complex of *Xenoturbella* contrasts with the reduction of the basal lamina in the Acoelomorpha.

The statocyst of the *Xenoturbella* shows no congruences at all with the gravity sensory organ of the Nemertodermatida or Acoela, and none with any of the other Plathelminthes, either.

Based on the original sperm pattern, *Xenoturbella* stands outside all of the Plathelminthomorpha. Further plesiomorphies such as the simple oral opening, the intestine without an anus, the intra-epidermal nervous system, and the lack of localized gonads should be mentioned, but have no meaning in the determination of kinship.

Xenoturbella cannot be assigned to the Spiralia either, as long as the cleavage pattern is unknown. Even the comparison of the epidermis of *Xenoturbella* with the skin of the Enteropneusta within the Radialia reveals only superficial similarities (PEDERSEN et al. 1988).

There is, in fact, only one apomorphous congruence which can be discussed in a justification of phylogenetic kinship relations, and that is the tip of epidermal cilia with only five peripheral double microtubules. The attempt to assess this identical shaft region as a synapomorphy of *Xenoturbella* and the Acoelomorpha, however, cannot be unified with an entire syndrome of strongly diverging features. The interpretation of this apomorphous congruence as convergence is forced a posteriori.

Currently, there are no arguments to convincingly hypothesize an adelphotaxon for *Xenoturbella bocki*. Its position in the phylogenetic system of the Eumetazoa is, therefore, still unclarified.

Euspiralia

■ **Autapomorphies**
(Fig. 55 → 3; 56 → 4)

– Intestine with anus.
  Anal pore in the ground pattern of the Euspiralia probably located dorsally just before the posterior end (BARTOLOMAEUS 1993b).

– Multiciliated epidermal cells.
  Convergent evolution to the multiciliation of the Plathelminthes (p. 140).

With the exception of the Plathelminthomorpha, the Spiralia have a "one-way intestine". Within the unity, we can employ the principle of parsimony, i.e., we can postulate a single evolution of an anus, and, based on this apomorphy, constitute a monophylum Euspiralia.

In the monophylum Plathelminthomorpha, the Gnathostomulida have the plesiomorphous state of monociliated epidermal cells, while the Plathelminthes exhibit the apomorphous state of multicili-ation (p. 133). This forces a posteriori the assumption that both multiciliated epidermal cells and multiciliated intestinal cells in the stem lineage of the Euspiralia evolved convergently.

Nemertini – Trochozoa

The Nemertini can clearly be justified as a monophylum on the basis of the unique proboscis and the rhynchocoel which surrounds it. In contrast to this, direct development, widespread among the Nemertini, must be assessed as an original state. The well-known pilidium larva appears only in a subgroup of the ribbon worms.

This brings us to the justification of the monophyly of the sister group Trochozoa via the trochophora larva (Fig. 99). A plankton larva, with a girdle of cilia (prototroch), which rings the body preorally, can be assumed to be part of the ground pattern of the following Spiralia unities – the Kamptozoa, Mollusca, and Articulata (realized in marine Annelida, including Sipunculida and Echiurida). We can postulate the unique evolution of a trochophora with a prototroch in the stem lineage of a subgroup of the Euspiralia which HATSCHEK (1878) named Trochozoa.

Nemertini

The ribbon worms are primarily inhabitants of the ocean floor. Numerous species varying in size from a few millimeters up to several meters in length inhabit widely differing sediments. In the ocean, the pelagic zone was also occupied (*Pelagonemertes*). Contact with other marine organisms led to commensals in bivalves (*Malacobdella*) and to parasites on crustaceans *(Carcinonemertes)*. A few species adapted to freshwater (*Prostoma*) and even to tropical soils (*Geonemertes*). We will begin with the characteristic features which prove the Nemertini to be a monophylum.

■ **Autapomorphies**
(Fig. 56 → 5)

– Proboscis as an organ for catching prey.
  Ectodermal infolding of the anterior end into the body. A retractor muscle fastens the proboscis in the rhynchocoel.

– Rhynchocoel (proboscis sheath) as a hydrostatic organ.
  The proboscis is wrapped in a fluid-filled cavity with its own epithelium. Rise in pressure through contraction of the peripheral musculature causes the eversion of the proboscis.

– Circulatory system with cellular, endothelial lining. Two lateral vessels, connected at the anterior and posterior ends by transverse vessels are part of the ground pattern.

– Ring-shaped central nervous system around the rhynchocoel.

Traditional Classification

A phylogenetic system of the Nemertini does not yet exist, but first contributions on the subject have already been made (CRANDALL 1993; MOORE and GIBSON 1993; SUNDBERG 1993).
In order to work out the ground pattern of the Nemertini, we have to refer back to the traditional classification into Anopla and Enopla, each with two high-ranking subunities (GIBSON 1972, 1982). The question as to the validity of these taxa will be examined in terms of an assessment of the diagnostic features as plesiomorphies (P) and apomorphies (A). Again here, it should be noted that the justification of individual unities as monophyla does not entail any statements on the sister group relationships within the Nemertini.

Anopla

Mouth and proboscis pore are separated from each other (P). Mouth beneath or behind the cerebral ganglia (P). Simple proboscis without a stylet (P). Longitudinal nerves peripherally located in the epidermis, beneath the epidermis or in the body wall musculature (P).

Based on these plesiomorphous features, the Anopla appear to be a paraphyletic taxon. It is possible, however, that certain rhabdoids in the epithelium of the proboscis form an autapomorphy of the Anopla (TURBEVILLE 1991). They are rod-shaped secretions a few μm in length with a filamentous axis and an electron dense medulla (Fig. 94D). Such rhabdoids are known only in the Palaeonemertini and the Heteronemertini, which brings us to the two subtaxa of the Anopla.

Palaeonemertini

Carinoma, Cephalothrix, Tubulanus, Carinina.
Body wall musculature with external circular muscles and internal longitudinal muscles (P), or with three layers with additional inner circular muscles. Longitudinal nerves outside the body wall musculature or in the layer of longitudinal muscles (P). Direct development (P). Probably a paraphyletic collection of relatively original Nemertini. Diagonally striated muscles have been observed in three species. This may be an autapomorphic feature (TURBEVILLE 1991).

Heteronemertini

Lineus, Cerebratulus.
Body wall musculature with four layers arranged from the outside to the inside as follows: weak circular muscles (TURBEVILLE and RUPPERT 1985; TURBEVILLE 1991), longitudinal muscles, circular muscles, and longitudinal muscles (A). Lateral nerves in the inner circular muscle layer (A). Indirect development through larvae with imaginal disks (A). Compared with the primarily double body wall musculature of the Palaeonemertini, the additional layers of circular and longitudinal muscles are an apomorphy. This is also true of the position of the lateral nerves in the inner circular muscle layer.

The monophyly of the Heteronemertini is convincingly justified by their mode of ontogenesis. It is only in this taxon that larvae with the differentiation of imaginal pockets exist.

Enopla

Double-layered body wall musculature composed of outer circular and inner longitudinal muscles (P). Oral opening in front of the brain (A). Lateral nerves internal to the body wall musculature (A). Direct development (P). The shifting of the mouth from the primary subterminal position as seen in the Anopla to the anterior end can be assessed as an autapomorphy of the Enopla. In comparison with the Heteronemertini, this is also true of the position of the nerve cords in the interior of the body.

Mouth and proboscis pore usually lead outwards through a common opening. This, however, is not an autapomorphy of the Enopla, since the mouth and proboscis remain separated from each other in some of the Hoplonemertini (Polystilifera).

Hoplonemertini

Amphiporus, Ototyphlonemertes, Carcinonemertes, Geonemertes, Prostoma, Pelagonemertes.
They have a stylet in the middle of the proboscis (A). Conventional division into the Monostilifera with one central stylet on a conical basis, and the Polystilifera with numerous small spines on a sickle-shaped pad.

Bdellomorpha

Malacobdella

Dorsoventrally flattened Nemertini with a sucker located ventrally at the posterior end (A). Commensals in the mantle cavity and between gills of marine bivalves (A). Lack of a stylet in the proboscis (assessment uncertain).

The unarmed proboscis may be the result of a reduction connected with the migration into bivalves and the conversion to microphagous nutrition. This argument would only be convincing, however, if the Bdellomorpha could be justified as a subtaxon of Hoplonemertini with a united mouth and proboscis pore. Opposing this is the possibility of an independent unification of the two openings and of a primary lack of an armed proboscis in the Bdellomorpha. This is supported by the fact that the esophagus of Malacobdella does not open into the rhynchodaeum (Monostilifera); on the contrary, the rhynchodaeum opens into an esophagus (CRANDALL 1993).

Ground Pattern and Evolutionary Alterations

Organization, ultrastructure, and development have been comprehensively described by GIBSON (1972), TURBEVILLE (1991) and FRIEDRICH (1979). Our statements on the ground pattern have to be based primarily on the allegedly paraphyletic Palaeonemertini. *Arhynchonemertes*, interpreted as a representative of the Nemertini lacking a proboscis (RISER 1988), cannot be taken into account until a more detailed structural analysis is undertaken.

Body Size – Habitus. We can postulate an organism measuring a few millimeters for the stem species of the Bilateria and the Plathelminthomorpha. Similar dimensions can be considered as part of the ground pattern of the Euspiralia and also part of the ground pattern of the Nemertini without conflict. The Palaeonemertini taxon *Cephalothrix* includes millimeter-sized species (GERNER 1969). A round body with conical anterior and posterior ends is also part of the ground pattern of the Nemertini.

Epidermis – Connective Tissue – Musculature. Multiciliated epidermal cells were adopted from the ground pattern of the Euspiralia. The cilia have a rostral rootlet and a vertical rootlet as well as a basal foot. Accessory centrioles are absent, as in the multiciliated epidermal cells of other Bilateria taxa. The cilia are surrounded by microvilli from the cell surface.

The epidermis has diverse glands. The production of slime and rod-shaped secretions may be a part of the trait pattern of the stem species of the Nemertini.

The connective tissue (parenchyma, mesenchyme) is composed of a fibrillar ECM and a number of cellular components. The ECM lies below the epidermis, embeds the musculature and surrounds all organs. It thus forms a continuous matrix skeleton in the body. This type of ECM certainly pertains to the ground pattern of the Nemertini. Detailed statements on the cellular components are, however, not possible. Corresponding to the small body size, the connective tissue cells in the stem species of the Nemertini as in the stem species of the Plathelminthomorpha were probably absent or only weakly developed. This is the case in smaller-sized species of *Cephalothrix* (GERNER 1969). Cells that could be neoblasts have not been identified with any degree of certainty.

A double-layered body wall musculature with an external circular musculature composed of a few layers of cells, and a thick inner layer of longitudinal muscles is a plesiomorphy in the ground pattern of the Nemertini. The commonly seen smooth musculature is an original trait.

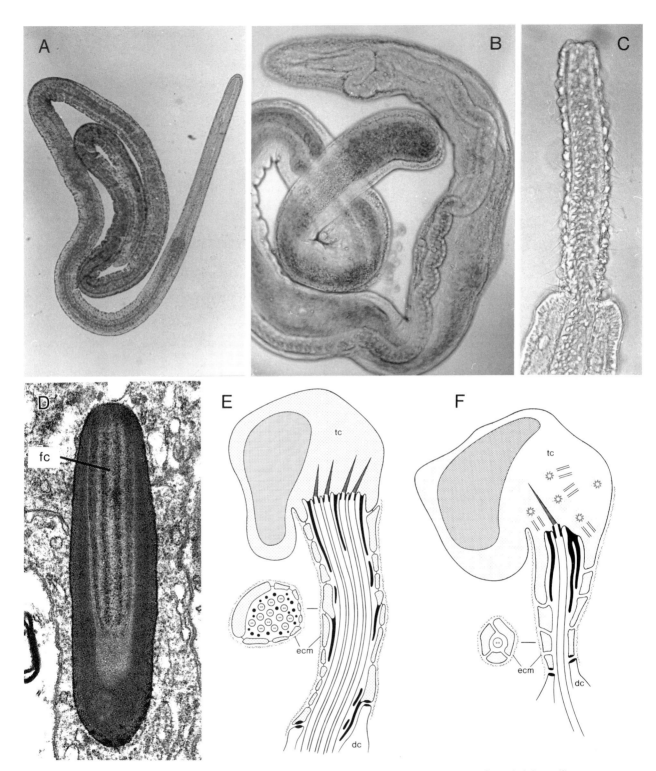

Fig. 94. Nemertini. A *Cephalothrix atlantica*. Habitus of the psammobiont species. Length 10 mm. B. *Cephalothrix pacifica* (Palaeonemertini). Slightly squeezed organism. Proboscis is visible inside the body. Length only 3–3.5 mm. C. *Cephalo-thrix pacifica*. Unarmed proboscis, partially everted. D. Rhabdoids of *Zygopleura rubens* (Heteronemertini). E. *Prostoma-tella arenicola* (Hoplonemertini). Protonephridium with multiciliary terminal cell. F. Stage during differentiation of the termi-nal cell of *Lineus viridis* (Heteronemertini) with one fully developed cilium and several centrioles. dc = canal cell. ecm = extracellular matrix. fc = filamentous axis. tc = terminal cell. (A-C Gerner 1969; D Turbeville 1991; E Bartolomaeus and Ax 1992; F Bartolomaeus 1985)

Intestine. There are no autapomorphies of the Nemertini that result from the structure of the digestive tract. In the ground pattern, the oral opening lies ventrally behind the tip of the body. It is primarily separated from the proboscis pore. The stomodaeum of ectodermal origin has multiciliated cells; the ultrastructure of the cilia is identical to that of cilia in epidermal cells. The adjacent mid-gut is primarily a simple, ciliated tube. The cilia are more loosely arranged than in the stomodaeum. The widespread lateral intestinal diverticula were not yet present in the small stem species of the Nemertini. The anus lies primarily dorsally, just before the posterior end (Fig. 95B).

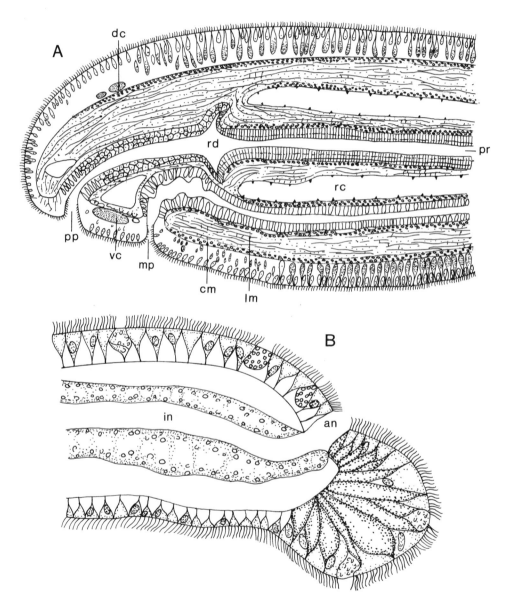

Fig. 95. Nemertini. A *Procarinina remanei* (Palaeonemertini). Sagittal section of the anterior end. Proboscis pore and mouth are separated (plesiomorphy). B. *Cephalothrix pacifica* (Palaeonemertini). Sagittal section of the posterior end with dorsal anus. an = anus. cm = circular muscles. dc = dorsal commissure. in = intestine. lm = longitudinal muscles. mp = mouth. pp = proboscis pore. pr = proboscis. rc = rhynchocoelom. rd = rhynchodaeum. vc = ventral commissure. (A Nawitzki 1931; B Gerner 1969)

Proboscis and Rhynchocoel. The functional unity composed of the proboscis as an ectodermal infolding of the anterior end and the fluid-filled rhynchocoel which surrounds the proboscis is the most prominent feature of the Nemertini. The prey-catching organ lies above the intestine. The out-folding of the proboscis is caused by the rise in fluid pressure in the rhynchocoel, which is effected by the contraction of the muscles surrounding the rhynchocoel. A retractor muscle connects the end of the proboscis with the wall of the rhynchocoel.

A proboscis, such as is realized in the Palaeonemertini and Heteronemertini, is part of the ground pattern. The proboscis pore and mouth are separated from each other (Fig. 95A). A short rhynchodaeum is followed by a long, closed tube, which is primarily undivided and unarmed. The epithelium, which contains many glands, is surrounded by muscles which can be interpreted as derivatives of the body wall musculature due to the ectodermal origin of the proboscis. The proboscis can be everted to a considerable length.

The armed proboscis of the Hoplonemertini first evolved within the Nemertini. Among the Polystilifera, the mouth and proboscis pore still remain separated. The opening of the mouth into the rhynchodaeum within the Monostilifera is more derived. In view of the considerable differences, a separate origin of the armed proboscis among the Polystilifera and Monostilifera is discussed (CRANDALL 1993).

We will describe the armed proboscis of the Monostilifera with its sharp division into three parts – an anterior proboscis cylinder, a short muscular bulb, and a posterior tubular section. The anterior portion of the muscular bulb is referred to as the diaphragm, which features stylets (Figs. 96,97). A conical stylet basis is deeply embedded in the diaphragm. The calcium phosphate stylet takes the form of a nail or a spiral composed of twisted strands. It protrudes freely into the anterior proboscis cylinder. A thin ductus ejaculatorius runs through the muscular bulb connecting the posterior and anterior parts.

While catching prey, the proboscis is everted only as far as the muscular bulb. The stylet causes wounds through which neurotoxins from the posterior proboscis cylinder enter the object of prey. The stylet has a very characteristic genesis. The basis and the stylet are produced separately. The basis is secreted in the diaphragm, whereas the stylets originate in the lateral stylet pockets. This applies to the first stylet that develops, as well as to the reserve stylets, which are replacements for the used weapon tips (STRICKER and CLONEY 1982). A large stylet formation cell, or styletocyte, is found in an epithelial lining of each of the usually paired stylet pockets. Several stylets are secreted intracellularly. Channels lead from the pockets to the anterior section of the proboscis. Differentiated stylets are transported through these channels towards the front and then mounted on the basis.

The rhynchocoel surrounds the proboscis on all sides. The cavity is lined with flat epithelial cells which overlie an ECM. Circular muscles and longitudinal muscles are on the outside.

Circulatory System. A circulatory system with an endothelial lining is a further characteristic. In the ground pattern, it consists of two lateral vessels located between the body wall and intestine, which are connected by a head and an anal vessel. This condition is seen in the taxon Cephalothricidae of the Palaeonemertini.

Manifold evolutionary changes arose within the Nemertini. Among the majority of the Anopla and in the Hoplonemertini, there is an additional unpaired dorsal vessel (Fig. 96A-C). Pseudometamerically arranged transversal loops connect the longitudinal vessels.

The vessels consist of a cellular endothelium, a subepithelial ECM, and a musculature consisting of circular and longitudinal muscles (Fig. 96D).

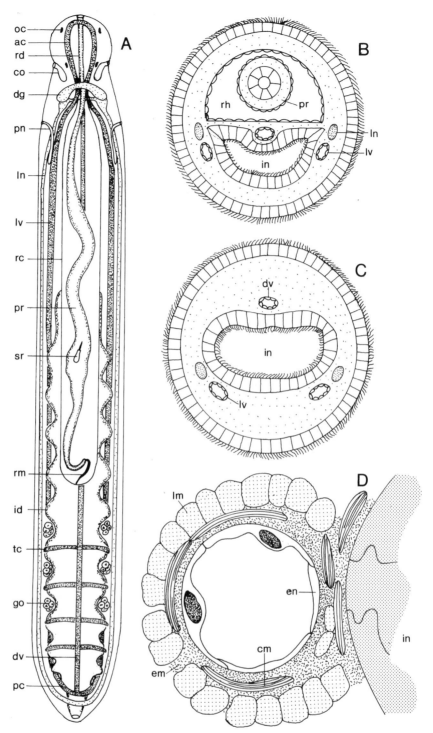

Fig. 96. Nemertini. A-C. Organization of the Hoplonemertini. A. Dorsal view. B. Cross section of the anterior body at the height of the proboscis and rhynchocoelom. C. Cross section of the posterior part behind the proboscis. D. Lateral blood vessel in Palaeonemertini based on studies of the ultrastructure. ac = anterior commissure of the circulatory system. cm = circular muscle. co = cerebral organ. dg = dorsal ganglion. dv = dorsal vessel. em = extracellular matrix. en = endothelium of the blood vessel. go = gonad. id = intestinal diverticulum. in = intestine. lm = longitudinal muscle. ln = lateral nerve. lv = lateral vessel. oc = ocellus. pc = posterior commissure of the circulatory system. pn = protonephridium. pr = proboscis. rc, rh = rhynchocoelom. rd = rhynchodaeum. rm = retractor muscle. sr = proboscis stylet. tc = transversal commissure of the circulatory system. (Turbeville 1991)

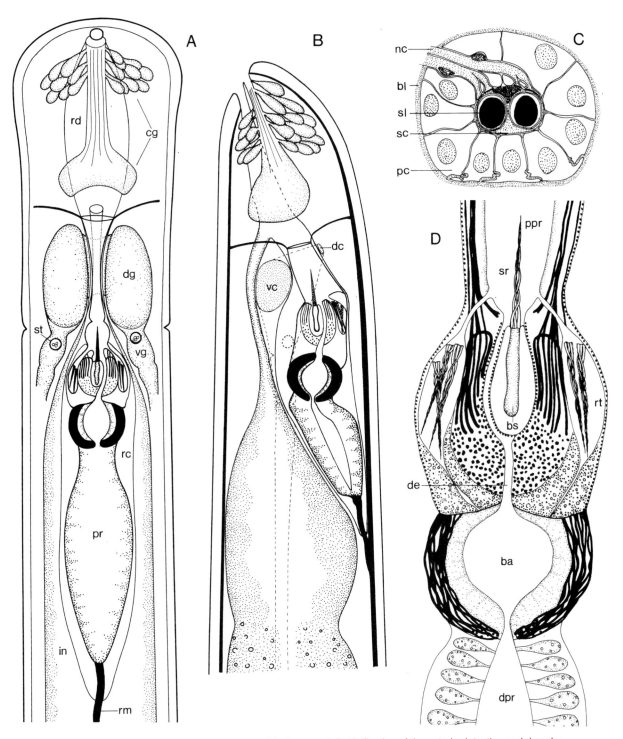

Fig. 97. Nemertini. Organization of *Ototyphlonemertes* (Hoplonemertini). Unification of the anterior intestine and rhyncho-daeum. Stylet apparature in the proboscis. A. *Ototyphlonemertes americana*. Horizontal view of the anterior end. B. *Ototyphlonemertes americana*. Sagittal section of the anterior end. C. *Ototyphlonemertes pallida*. Sagittal section of the stato-cyst. D. *Ototyphonemertes americana*. Stylet apparatus of the proboscis. Horizontal view. ba = balloon. bl = basal lamina. bs = stylet basis. cg = cerebral glands. dc = dorsal commissure. de = ductus ejaculatorius. dg = dorsal ganglion. dpr = distal part of the proboscis. in = intestine. nc = nerve cell. pc = parietal cell. ppr = proximal part of the proboscis. pr = pro-boscis. rc = rhynchocoelom. rd = rhynchodaeum. rm = retractor muscle. rt = pocket for reserve stylets. sc = statolith cham-ber cell. sl = statolith. sr = proboscis stylet. st = statocyst. vc = ventral commissure. vg = ventral ganglion. (A, B, D Gerner 1969; C Brüggemann and Ehlers 1981)

Protonephridia. Paired protonephridia with the following elements are to be included in the ground pattern of the Nemertini. Each protonephridium has several terminal cells, several canal cells and a nephropore in the epidermis of the anterior part of the body. The terminal cells are multiciliated, and each cilium has one rootlet. Microvilli arise between the cilia (Fig. 94E). The terminal cells are elongated distally to form a cytoplasmic, hollow cylinder with numerous slits. The filter region is covered with ECM (BARTOLOMAEUS 1988; BARTOLOMAEUS and AX 1992).

The protonephridia lie freely in the extracellular matrix of the compact body. Close contact with the circulatory system is widely established within the Nemertini. Terminal cells are an integral component of the vessel wall in *Tubulanus annulatus* (JESPERSEN and LÜTZEN 1987). The terminal cells, however, are always separated from the vessel lumen by wall cells with ECM, sometimes by ECM alone (TURBEVILLE 1991).

Multiciliarity is an apomorphy in contrast to the monociliated terminal cells in the Gnathostomulida or Gastrotricha. The development of the protonephridia in *Lineus viridis* passes through the stage of a monociliated terminal cell (Fig. 94F; BARTOLOMAEUS 1985).

Nervous System – Sensory Organs. One pair of dorsal and one pair of ventral cerebral ganglia, one pair of lateral longitudinal nerves and a peripheral nerve net belong to the ground pattern. The cerebral ganglia are connected by commissures to a ring around the rhynchodaeum or anterior rhynchocoel. The longitudinal nerves arise from the ventral ganglia. Cerebral ganglia and lateral nerves are found primarily in the epidermis (Palaeonemertini: *Carinina*). Additional dorsal and ventral longitudinal nerves evolved within the Nemertini. Among the remaining Bilateria, ring-shaped brains, if developed, are found around the stomodaeum. The position in the Nemertini must be assessed as an autapomorphy, which originated in connection with the evolution of the proboscis.

Sensory organs such as a frontal organ at the anterior end, photoreceptors as subepidermal pigment cup cells or the tube-shaped ectodermal cerebral organ (Fig. 96A) are intermittently prevalent. Incontestable statements about them in the ground pattern do not appear to be possible.

Statocysts offer a prime example of the evolution of a sensory organ within the Nemertini. Statocysts are realized in *Ototyphlonemertes*, a taxon of psammobiontic Hoplonemertini. The paired gravity sensory organs are found on the elongated ventral ganglia (Fig. 97A,C). There are considerable ultrastructural differences in comparison with the statocysts in the Plathelminthes or other metazoan taxa. *Ototyphlonemertes pallida* has a statolith chamber cell in the center of the statocyst, which contains on an average three statoliths in separate chambers. Several nerve cells lead from the dorsal surface of the chamber cell out of the statocyst. The statolith chamber cell and nerve cord are surrounded by numerous parietal cells.

The statocyst is an autapomorphy of the taxon *Ototyphlonemertes* (BRÜGGEMANN and EHLERS 1981). In other words, it evolved in the stem lineage of a small monophylum of the Nemertini, 20 species of which are known today (NORENBURG 1988).

Gonads – Gametes – Development. The sexual organization shows simple, predominantly plesiomorphous traits. The Nemertini are primarily gonochoric, with only isolated cases of a change to hermaphroditism.

Entolecithal egg production and an original sperm pattern with acrosome, perforatorium, four mitochondria, and one cilium form part of the ground pattern. External fertilization is correlated with this. Only the large number of gonads and their arrangement in two rows can be an apomorphy in the ground pattern of the Nemertini. Simple gonoducts develop at the time of maturation from the gonads leading to the body surface.

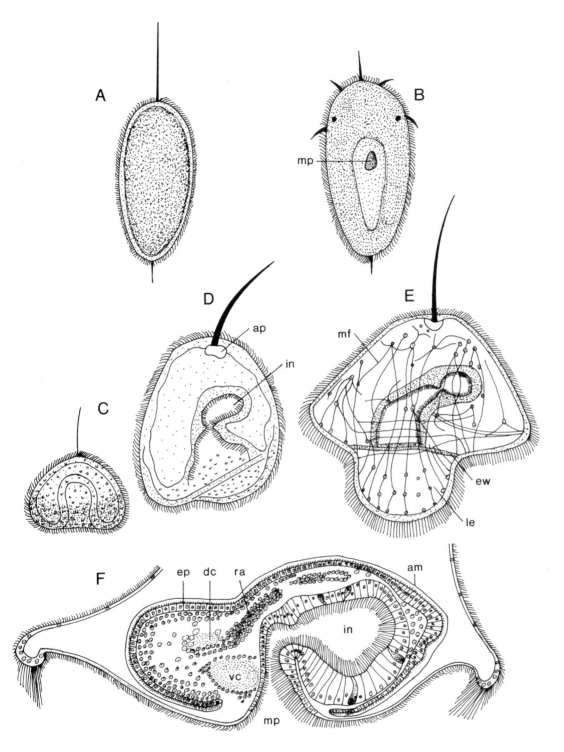

Fig. 98. A, B. Direct development through free-swimming juvenile stages in Palaeonemertini. A. *Tubulanus punctatus.* B. *Procephalothrix simulus.* C-F. Indirect development through the pilidium larva in the Heteronemertini. C-E. *Lineus torquatus.* Differentiation from gastrula to pilidium. F. Sagittal section of the ventral portion of a pilidium larva. Development of the adult in the amniotic cavity of the larva. am = amnion. ap = apical plate. dc = dorsal commissure. ep = epithelium of the adult. ew = epithelial bulge. in = intestine. le = lateral lobes. mf = muscle fiber. mp = mouth. ra = proboscis anlage. vc = ventral commissure. (Friedrich 1979)

The Palaeonemertini, Hoplonemertini, and Bdellomorpha develop directly with or without a pelagic juvenile stage. An apical organ with a tuft of cilia is seen. There are, however, no other larval features (Fig. 98A,B).

The planktotrophic pilidium larva is only realized in the taxon Heteronemertini (Fig. 98C-F). A bell-shaped, completely ciliated body with an apical organ develops from the gastrula. The margins next to the mouth then grow to form lobes resembling a fencing helmet. Characteristic of the pilidium are the imaginal disks. Several swellings differentiate themselves in the larval ectoderm – one pair of head, cerebral, and trunk disks as well as an unpaired dorsal disk. The imaginal discs sink inside the body and form pockets, which grow together to surround the larval intestine. The epidermis of the adult originates from the inner leaves of the vesicles. The outer leaves unite to form an amnion, which surrounds the central part of the larva in an amniotic cavity. The adult originates only from this part. The rest of the larva is discarded.

Pelagic lecithotrophous larvae (*Micrura*) and intracapsular larvae in egg capsules (*Lineus* species) represent evolutionary alterations of the pilidium larva within the Heteronemertini. The connection is documented by the imaginal disks characteristic of all larvae.

A pilidium larva can be ruled out as a feature of the ground pattern of the Nemertini. Were this not so, the direct development would have to be a derived state in the Palaeonemertini, Hoplonemertini and Bdellomorpha. There is, however, nothing to support this hypothesis. The pilidium is a characteristic autapomorphy of the subtaxon Heteronemertini.

This means that the widespread attempt to homologize the pilidium larva with the trochophora larva of the Trochozoa cannot succeed. In addition, the existence of imaginal disks in the pilidium marks a characteristic, incomparable path of development.

Evolutionary Assessment of the Rhynchocoel and Circulatory System.

With their cellular epithelial lining, the rhynchocoel and circulatory system fulfil the definition of the "secondary body cavity" (p. 116). This alone, however, does not justify the conclusion of a homology with homonymous body cavities in other Spiralia. The prerequisite for a homologization of rhynchocoel and circulatory system of the Nemertini with the coelom of the Annelida would be a well-founded concept of a body cavity condition in a stem species from which the body cavities of both unities are derivable. Such a concept does not exist. There are no arguments to assume the evolution of the rhynchocoel, as the motor for the eversion of the proboscis, and of the circulatory system of the Nemertini from a polymerous hydroskeleton with locomotory function or vice versa. The rhynchocoel and circulatory system are to be assessed as genuine features of ribbon worms, which evolved in the stem lineage of the Nemertini independently of secondary body cavities of other Spiralia.

Summary of Features of the Ground Pattern

The following features are assigned to the stem species of the Nemertini (P = plesiomorphy, A = apomorphy).

− Small round body (P).

− Multiciliated epidermis (P).

− Compact body without ore with only weakly developed connective tissue (P).

− Double-layered body wall musculature composed of outer circular muscles and inner longitudinal muscles (P).

210

- Smooth musculature (P).

- Subterminal oral opening, separated from the proboscis pore (P). Multiciliated intestinal cells (P). Dorsal anus positioned just before the posterior end (P).

- Proboscis with rhynchodaeum and an undivided, unarmed tube (A).

- Rhynchocoel as an epithelially lined cavity around the proboscis (A).

- Circulatory system with endothelial lining (A).

- Paired protonephridia in the ECM of the compact body (P).

- Four cerebral ganglia in a ring around the proboscis (A). Two lateral longitudinal nerves in the epidermis (P).

- Gonochoric (P). Entolecithal egg production (P). Primary sperm pattern (P). External fertilization (P). Numerous gonads arranged in longitudinal rows (A).

- Direct development without larva (P).

Trochozoa

■ **Autapomorphy**
(Fig. 56 → 6)

- Trochophora larva
  With a prototroch, a pair of ocelli, and (?) an apical organ.

A planktonic larva can, with the exception of the Nemertini, be postulated as a feature of the ground pattern of the other high-ranking subtaxa of the Euspiralia (p. 200). A biphasic life cycle with a plankton larva is therefore postulated as an autapomorphy of the unity Trochozoa comprising them.
The structure of the trochophore larva can be specified by the presence of three features (Fig. 99).

Prototroch
The larvae of the Trochozoa are characterized by a girdle of cilia which circles the body preorally, and divides the body into an episphere and hyposphere. The prototroch is a characteristic feature in the organization of the larva. Accordingly, we take the term trochophora for this larva in the ground pattern of the Trochozoa.

Ocelli
Photosensitive organs composed of a few rhabdomeric receptor cells and pigment cells are hypothesized for the trochophora larva of the stem species of the Mollusca and of the Annelida (BARTOLOMAEUS 1987, 1992a,b). In addition to this, paired ocelli of ectodermal origin are to be postulated as a derived feature of the trochophora of all Trochozoa (BARTOLOMAEUS).

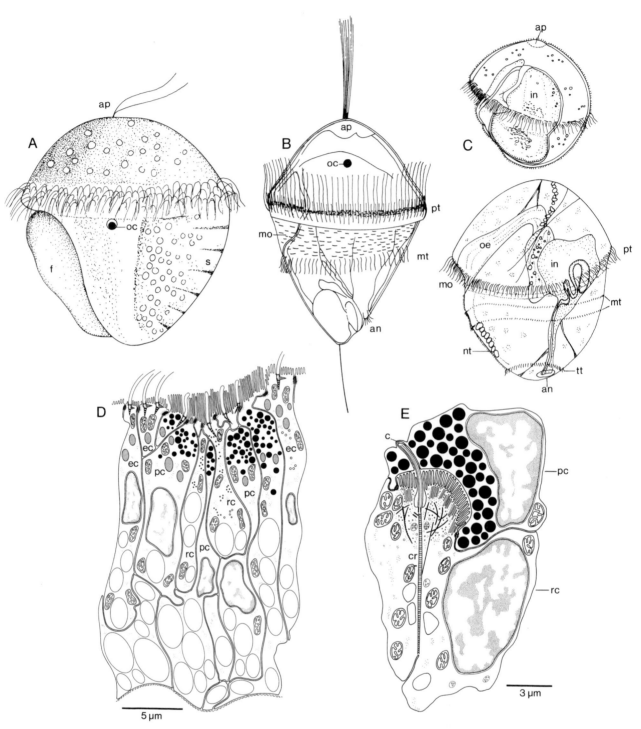

Fig. 99. Trochozoa. A-C. Side view of trochopora larvae of different taxa. A. *Ischnochiton magdalensis* (Mollusca, Polyplacophora). B. *Pomatoceros* (Annelida). C. *Echiurus abyssalis* (Echiurida). Young and older larvae. D. Intraepidermal photoreceptor of *Lepidochiton cinereus* larva (Mollusca). The everse eyes are composed of four elongated receptor cells and six highly prismatic pigment cells; both form of cells have two to three cilia. E. Subepidermal photoreceptor of *Anaitides mucosa* larva (Annelida). The inverse eyes are composed of a cup-shaped pigment cell and a rhabdomeric receptor cell. The latter has a short cilium, an accessory centriole and a striated ciliary rootlet. a = anus. ap = apical organ. c = cilium. cr = ciliary rootlet. ec = ectodermal cell. in = intestine. mo = mouth. mt = metatroch. nt = neurotrochoid. oc = ocellus. oe = esophagus. pc = pigment cell. pt = prototroch. rc = receptor cell. s = shell anlage. tt = telotroch. (A Heath, in Hoffmann 1930; B de Beauchamp 1961a; C Korn 1960; D Bartolomaeus 1992b; E Bartolomeus 1987)

Apical Organ

An apical organ with a tuft of long cilia is the third characteristic feature of the trochophora larva. However, a comparable organ also appears in the development of Nemertini with and without larva (p. 210). In contrast to the prototroch and ocelli, the following interpretation is possible. An apical organ was already realized in the stem species of the Euspiralia, which developed directly. During the evolution of the trochophora, it was adopted in the stem lineage of the Trochozoa as a plesiomorphy in the trochophora. However, there are considerable ultrastructural differences between the Nemertini and Annelida. In the pilidium larvae of the Heteronemertini, the apical disk has monociliated or multiciliated collar cells. The cilia here are surrounded by a wreath of numerous microvilli (CANTELL et al. 1982). Comparable structures in the apical organ of the annelid trochophora are unknown (VERGER-BOCQUET 1992).

A unity composed of Kamptozoa + Mollusca and the Articulata probably form the highest ranking adelphotaxa of the Trochozoa (BARTOLOMAEUS 1993b). They will be treated in the second volume of this book series.

Literature

AFZELIUS, B. A. & M. FERRAGUTI (1978). The spermatozoon of Priapulus caudatus Lamarck. J. Submicrosc. Cytol. **10**, 71–80.

ALLMAN, G. J. (1853). Anatomy and physiology of Cordylophora. Phil. Trans. London.

AX, P. (1952). Ciliopharyngiella intermedia nov. gen. nov. spec., Repräsentant einer neuen Turbellarien-Familie des marinen Mesopsammon. Zool. Jb. Syst. **81**, 285–312.

AX, P. (1954). Marine Turbellaria Dalyellioida von den deutschen Küsten. I. Die Gattungen Baicalellia, Hangethellia und Canetellia. Zool. Jb. Syst. **82**, 481–496.

AX, P. (1956a). Monographie der Otoplanidae (Turbellaria). Morphologie und Systematik. Akad. d. Wiss. u. d. Lit. Mainz. Abhandl. d. Math.-Naturw. Kl. Jg. 1955, **13**, 1–298.

AX, P. (1956b). Die Gnathostomulida, eine rätselhafte Wurmgruppe aus dem Meeressand. Akad. d. Wiss. u. d. Lit. Mainz. Abhandl. d. Math.-Naturw. Kl. Jg. 1956, **8**, 1–32.

AX, P. (1956c). Les Turbellariés des Etangs Côtiers du Littoral Méditerranéen de la France Méridionale. Vie et Milieu, Suppl. **5**, 1–215.

AX, P. (1963). Relationships and Phylogeny of the Turbellaria. In E. C. DOUGHERTY (Ed.). The lower Metazoa. 191–224. University of California Press.

AX, P. (1964a). Das Hautgeißelepithel der Gnathostomulida. Verh. Deutsch. Zool. Ges. München 1963, 452–461.

AX, P. (1964b). Die Kieferapparatur von Gnathostomaria lutheri Ax (Gnathostomulida). Zool. Anz. **173**, 174–181.

AX, P. (1965). Zur Morphologie und Systematik der Gnathostomulida. Untersuchungen an Gnathostomula paradoxa Ax. Z. zool. Syst. Evolut.-forsch. **3**, 259–276.

AX, P. (1966a). Die Bedeutung der interstitiellen Sandfauna für allgemeine Probleme der Systematik, Ökologie und Biologie. Veröffentl. Inst. Meeresf. Bremerhaven, Sonderband II, 15–66.

AX, P. (1966b). Eine neue Tierklasse aus dem Litoral des Meeres – Gnathostomulida. Umschau 1966, 17–23.

AX, P. (1969). Populationsdynamik, Lebenszyklen und Fortpflanzungsbiologie der Mikrofauna des Meeressandes. Verh. Deutsch. Zool. Ges. Innsbruck 1968, 66–113.

AX, P. (1971). Zur Systematik und Phylogenie der Trigonostominae (Turbellaria, Neorhabdocoela). Mikrofauna Meeresboden **4**, 1–84.

AX, P. (1984). Das Phylogenetische System. Systematisierung der lebenden Natur aufgrund ihrer Phylogenese. G. Fischer. Stuttgart, New York.

AX, P. (1985). Stem species and the stem lineage concept. Cladistics **1**, 279–287.

AX, P. (1987). The phylogenetic system. The systematization of organisms on the basis of their phylogenesis. J. Wiley & Sons. Chichester.

AX, P. (1988). Systematik in der Biologie. Darstellung der stammesgeschichtlichen Ordnung in der lebenden Natur. UTB 1502. G. Fischer, Stuttgart.

AX, P. (1989a). Homologie in der Biologie – ein Relationsbegriff im Vergleich von Arten. Zool. Beitr. N. F. **32**, 487–496.

AX, P. (1989b). The integration of fossils in the phylogenetic system of organisms. Abh. naturwiss.Ver.Hamburg (NF) **28**, 27–43.

AX, P. (1989c). Basic phylogenetic systematization of the Metazoa. In B. FERNHOLM, K. BREMER & H. JÖRNVALL (Eds.). The hierarchy of life. Nobel. Symp. **70**, 229–245. Elsevier Science Publishers, Amsterdam.

AX, P. & R. AX (1967). Turbellaria Proseriata von der Pazifikküste der USA (Washington). I. Otoplanidae. Z. Morph. Tiere **61**, 215–264.

AX, P. & J. DÖRJES (1966). Oligochoerus limnophilus nov. spec., ein kaspisches Faunenelement als erster Süßwasservertreter der Turbellaria Acoela in Flüssen Mitteleuropas. Int. Rev. ges. Hydrobiol. **57**, 15–44.

AX, P. & E. SCHULZ (1959). Ungeschlechtliche Fortpflanzung durch Paratomie bei acoelen Turbellarien. Biol. Zentralbl. **78**, 613–621.

AX, P., SOPOTT-EHLERS, B., EHLERS, U. & T. BARTOLOMAEUS (1989). Was leistet das Elektronenmikroskop für die Aufdeckung der Stammgeschichte der Tiere? Akademie der Wissenschaften und der Literatur Mainz 1949–1989, 73–86. F. Steiner. Wiesbaden, Stuttgart.

BACCETTI, B. (1979). The evolution of the acrosomal complex. In D. W. FAWCETT & J. M. BEDFORD (Eds.). The Spermatozoon. Maturation, motility, surface properties and comparative aspects. 305–329. Urban & Schwarzenberg. Baltimore, Munich.

BAGUÑÀ, J., SALO, E. & R. ROMERO (1989). Evidence that neoblasts are totipotent stem cells and the source of blastema cells. Development **107**, 77–86.

BARNES, J. (1985). The complete works of Aristotle. Vol. 1 u 2. Princeton University Press. Princeton, New Jersey.

BARTOLOMAEUS, T. (1985). Ultrastructure and development of the protonephridia of Lineus viridis (Nemertini). Microfauna Marina **2**, 61–83.

BARTOLOMAEUS, T. (1987). Ultrastruktur des Photorezeptors der Trochophora von Anaitides mucosa Oersted (Phyllodocidae, Annelida). Microfauna Marina **3**, 411–418.

BARTOLOMAEUS, T. (1988). No direct contact between the excretory system and the circulatory system in Prostomatella arenicola Friedrich (Hoplonemertini). Hydrobiologia **156**, 175–181.

BARTOLOMAEUS, T. (1992a). Ultrastructure of the photoreceptors in certain larvae of the Annelida. Microfauna Marina **7**, 191–214.

BARTOLOMAEUS, T. (1992b). Ultrastructure of the photoreceptor in the larvae of Lepidochiton cinereus (Mollusca, Polyplacophora) and Lacuna divaricata (Mollusca, Gastropoda). Microfauna Marina **7**, 215–236.

BARTOLOMAEUS, T. (1993a). Die Leibeshöhlenverhältnisse und Nephridialorgane der Bilateria – Ultrastruktur, Entwicklung und Evolution. Habilitationsschrift Univ. Göttingen.

BARTOLOMAEUS, T. (1993b). Die Leibeshöhlenverhältnisse und Verwandtschaftsbeziehungen der Spiralia. Verh. Deutsch. Zool. Ges. **86**, 1. 42.

BARTOLOMAEUS, T. (1994). On the ultrastructure of the coelomic lining in the Annelida, Sipuncula and Echiura. Microfauna Marina **9**, 171–220.

BARTOLOMAEUS, T. & P. AX (1992). Protonephridia and Metanephridia – their relation within the Bilateria. Z. zool. Syst. Evolut.-forsch. **30**, 21–45.

BAYER, F. M. & H. B. OWRE (1968). The free-living lower Invertebrates. The MacMillan Company, New York.

BEAUCHAMP, P. de (1961a). Généralités sur les Métazoires triploblastiques. Traité de Zoologie **4** (1), 1–19.

BEAUCHAMP, P. de (1961b). Classe des Turbellariés. Traité de Zoologie **4** (1), 35–212.

BOCK, S. (1913). Studien über Polycladen. Zool. Bidrag Uppsala **2**, 31–344.

BÖGER, H. (1988). Versuch über das phylogenetische System der Porifera. Meyniana **40**, 143–154.

BORKOTT, H. (1970). Geschlechtliche Organisation, Fortpflanzungsverhalten und Ursachen der sexuellen Vermehrung von Stenostomum sthenum nov. spec. Z. Morph. Tiere **67**, 183–262.

BOUILLON, J. (1983). Sur le cycle biologique de Eirene hexanemalis (Goette 1886) (Eirenidae, Leptomedusae, Hydrozoa, Cnidaria). Cah. Biol. Mar. **24**, 421–427.

BOUILLON, J. (1985). Essai de classification des Hydropolypes-Hydroméduses (Hydrozoa-Cnidaria). Indo-Malayan Zoology **1**, 29–243.

BOUILLON, J. (1987). Considérations sur le développement des Narcoméduses et sur leur position phylogénétique. Indo-Malayan Zoology **4**, 189–278.

BRESCIANI, J. & T. FENCHEL (1965). Studies on dicyemid Mesozoa. I. The fine structure of the adult (The Nematogen and Rhombogen stage). Vidensk. Medd. fra Dansk. naturh. Foren. **128**, 85–92.

BRESCIANI, J. & T. FENCHEL (1967). Studies on dicyemid Mesozoa. II. The fine structure of the infusoriform larva. Ophelia **4**, 1–18.

BRESSLAU, E. (1928–33). Turbellaria. Hand. d. Zool. **2** (1), 52–304.

BRIDGE, D., SCHIERWATER, B., CUNNINGHAM, C. W., DE SALLE, R. & L. W. BUSS (1992). Mitochondrial DNA structure and the phylogenetic relationships of recent cnidaria classes. Proc. Natl. Acad. USA **89**, 8750–8753.

BRÜGGEMANN, J. & U. EHLERS (1981). Ultrastruktur der Statocyste von Ototyphlonemertes pallida (Keferstein, 1887) (Nemertini). Zoomorphology **97**, 75–87.

BRUSCA, R. C. & G. J. BRUSCA (1990). Invertebrates. Sinauer Associates, Inc. Sunderland, Massachusetts.

BUNGE, M. (1977). Treatise on basic philosophy. Vol. 3. Ontology I: The furniture of the world. D. Reidel. Dordrecht, Boston.

BUNGE, M. (1979). Treatise on basic philosophy. Vol. 4. Ontology II: A world of systems. D. Reidel. Dordrecht, Boston, London.

CANTELL, C.-E., FRANZÉN, A. & T. SENSENBAUGH (1982). Ultrastructure of multiciliated collar cells in the pilidium larva of Lineus bilineatus (Nemertini). Zoomorphology **101**, 1–15.

CHEVALIER, J.-P. (1987). Ordre des Scléractiniaires. Traité de Zoologie **3** (3), 403–764.

CHRISTEN, R. (1994). Molecular phylogeny and the origin of Metazoa. In BENGTSON, S. (Ed.). Early life on earth. Nobel Symp. **84**, 467–474. Columbia University Press, New York.

CHRISTEN, R., RATTO, A., BAROIN, A., PERASSO, R., GRELL, K. G. & A. ADOUTTE (1991). Origin of metazoans. A phylogeny deduced from sequences of 28S ribosomal RNA. In A. M. SIMONETTA & S. CONWAY MORRIS (Eds.). The early evolution of Metazoa and the significance of problematic taxa. 1–9. Cambridge University Press, Cambridge.

COIL, W. H. (1991). Platyhelminthes: Cestoidea. In F. W. HARRISON & B. J. BOGITSH (Eds.). Microscopic Anatomy of Invertebrates **3**, 211–283. Wiley-Liss, Inc. New York.

COPELAND, H. F. (1934). The kingdoms of organisms. Quart. Rev. Biol. **13**, 383–420.

CRANDALL, F. B. (1993). Major characters and enoplan systematics. Hydrobiologia **266**, 115–140.

DOE, D. A. (1981). Comparative ultrastructure of the pharyx simplex in Turbellaria. Zoomorphology **97**, 133–192.

DOHLE, W. (1989). Zur Frage der Homologie ontogenetischer Muster. Zool. Beitr. N. F. **32**, 355–389.

DÖNGES, J. (1988). Parasitologie. G. Thieme. Stuttgart, New York.

EERNISSE, D. J., ALBERT, J. S. & F. E. ANDERSON (1992). Annelida and Arthropoda are not sister taxa: A phylogenetic analysis of spiralian metazoan morphology. Syst. Biol. **41**, 305–330.

EHLERS, U. (1981). Fine structure of the giant aflagellate spermatozoon in Pseudostomum quadrioculatum (Leuckart) (Platyhelminthes, Prolecithophora). Hydrobiologia **84**, 287–300.

EHLERS, U. (1985). Das phylogenetische System der Platyhelminthes. G. Fischer. Stuttgart, New York.

EHLERS, U. (1986). Comments on a phylogenetic system of the Plathelminthes. Hydrobiologia **132**, 1–12.

EHLERS, U. (1988). The Prolecithophora – a monophyletic taxon of the Plathelminthes? Fortschritte der Zoologie/Progress in Zoology **36**, 359–365.

EHLERS, U. (1991). Comparative morphology of statocysts in the Plathelminthes and the Xenoturbellida. Hydrobiologia **227**, 263–271.

EHLERS, U. (1992a). On the fine structure of Paratomella rubra Rieger & Ott (Acoela) and the position of the taxon Paratomella Dörjes in a phylogenetic system of the Acoelomorpha (Plathelminthes). Microfauna Marina **7**, 265–293.

EHLERS, U. (1992b). No mitosis of differentiated epidermal cells in Rhynchoscolex simplex Leidy, 1851 (Catenulida). Microfauna Marina **7**, 311–321.

EHLERS, U. (1992c). Frontal glandular and sensory structures in Nemertoderma (Nemertodermatida) and Paratomella (Acoela): Ultrastructure and phylogenetic implications for the monophyly of the Euplathelminthes (Plathelminthes). Zoomorphology **112**, 227–236.

EHLERS, U. (1993). Ultrastructure of the spermatozoa of Halammohydra schulzei (Cnidaria, Hydrozoa): the significance of acrosomal structures for the systematization of the Eumetazoa. Microfauna Marina **8**, 115–130.

EHLERS, U. (1994a). On the ultrastructure of the protonephridium of Rhynchoscolex simplex and the basic systematization of the Catenulida (Plathelminthes). Microfauna Marina **9**,157–169.

EHLERS, U. (1994b). Absence of a pseudocoel or pseudocoelom in Anoplostoma vivipara (Nematodes). Microfauna Marina **9**, 345–350.

EHLERS, U. (1995). The basic organization of the Plathelminthes. Hydrobiologia **305**, 21–26.

FAUBEL, A. (1974). Macrostomida (Turbellaria) von einem Sandstrand der Nordseeinsel Sylt. Mikrofauna Meeresboden **45**, 1–32.

FAUBEL, A. (1983). The Polycladida, Turbellaria. Proposal and establishment of a new system. Part I. The Acotylea. Mitt. hamb. zool. Mus. Inst. **80**, 17–121.

FAUBEL, A. (1984). The Polycladida, Turbellaria. Proposal and establishment of a new system. Part II. The Cotylea. Mitt. hamb. zool. Mus. Inst. **81**, 189–259.

FELL, P. E. (1989). Porifera. In K. G. & R. G. ADIYODI (Eds.). Reproductive Biology of Invertebrates **4**, A, 1–41.

FIELD, K. G., OLSEN, G. J., LANE, D. J., GIOVANNONI, S. J., GHISELIN, M. T., RAFF, E. C., PACE, N. R. & R. A. RAFF (1988). Molecular phylogeny of the animal kingdom. Science **239**, 748–753.

FRANSEN, M. E. (1988). Coelomic and vascular systems. In W. WESTHEIDE & C. O. HERMANS (Eds.). The Ultrastructure of Polychaeta. Microfauna Marina **4**, 199–213. G. Fischer. Stuttgart, New York.

FRANZÉN, A. (1956). On spermiogenesis, morphology of the spermatozoon, and biology of fertilization among Invertebrates. Zool. Bidr. Uppsala **31**, 355–482.

FRANZÉN, A. & B. A. AFZELIUS (1987). The ciliated epidermis of Xenoturbella bocki (Platyhelminthes, Xenoturbellida) with some phylogenetic considerations. Zool.Scripta **16**, 9–17.

FRIED, G. & M. A. HASEEB (1991). Platyhelminthes: Aspidogastrea, Monogenea, and Digenea. In F. W. HARRISON & B. S. BOGITSH (Eds.). Microscopic Anatomy of Invertebrates **3**, 141–209, Wiley-Liss, Inc. New York.

FRIEDRICH, H. (1979). Nemertini. In F. Seidel (Ed.): Morphogenese der Tiere. Erste Reihe. Lieferung 3: D5, 1–136. G. Fischer. Stuttgart, New York.

GALLISSIAN, M.-F. & J. VACELET (1992). Ultrastructure of the oocyte and embryo of the calcified sponge, Petrobiona massiliana (Porifera, Calcarea). Zoomorphology **112**, 133–141.

GARDINER, B. G. (1993). Haematothermia: warm-blooded Amniotes. Cladistics **9**, 369–395.

GARDINER, S. L. & R. Rieger (1980). Rudimentary cilia in muscle cells of annelids and echinoderms. Cell. Tissue Res. **213**, 247–252.

GERNER, L. (1969). Nemertinen der Gattungen Cephalothrix und Ototyphlonemertes aus dem marinen Mesopsammal. Helgoländer wiss. Meeresunters. **19**, 68–110.

GHISELIN, M. T. (1969). The evolution of hermaphroditism among animals. Quart. Rev. Biol. **44**, 189–208.

GHISELIN, M. T. (1974a). A radical solution of the species problem. Syst. Zool. **23**, 536–544.

GHISELIN, M. T. (1974b). The economy of nature and the evolution of sex. University of California Press. Berkeley. Los Angeles, London.

GIBSON, R. (1972). Nemerteans. Hutchinson University Library, London.

GIBSON, R. (1982). Nemertea. In S. B. Parker (Ed.). Synopsis and classification of living organisms. 823–846. McGraw-Hill Book Company, New York.

GIESA, S. (1966). Die Embryonalentwicklung von Monocelis fusca (Turbellaria, Proseriata). Z. Morph. Ökol. Tiere **57**, 137–230.

GRELL, K. G. (1971). Trichoplax adhaerens F. E. Schulze und die Entstehung der Metazoen. Naturw. Rundschau **24**, 160–161.

GRELL, K. G. (1972). Eibildung und Furchung von Trichoplax adhaerens F. E. Schulze (Placozoa). Z. Morphol. Tiere **73**, 297–314.

GRELL, K. G. (1981). Trichoplax adhaerens and the origin of Metazoa. In: Origine dei Grandi Phyla dei Metazoi. Acc. Naz. Lincei. Atti dei Convegni Lincei **49**, 107–121.

GRELL, K. G. & G. BENWITZ (1974). Elektronenmikroskopische Beobachtungen über das Wachstum der Eizelle und die Bildung der „Befruchtungsmembran" von Trichoplax adhaerens F. E. Schulze (Placozoa). Z. Morph. Tiere **79**, 295–310.

GRELL, K. G. & G. BENWITZ (1981). Ergänzende Untersuchungen zur Ultrastruktur von Trichoplax adhaerens F. E. Schulze (Placozoa). Zoomorphology **98**, 47–67.

GRELL, K. G. & A. RUTHMANN (1991). Placozoa. In F. W. HARRISON & J. A. WESTFALL (Eds.). Microscopic Anatomy of Invertebrates **2**, 13–27. Wiley-Liss, Inc. New York.

GROBBEN, C. (1908). Die systematische Einteilung des Tierreichs. Verh. zool.-bot. Ges. Wien **58**, 491–501.

HAECKEL, E. (1866). Generelle Morphologie der Organismen. Band I und II. G. Reiner, Berlin.

HAECKEL, E. (1874). Die Gastraea-Theorie, die phylogenetische Classification des Thierreichs und die Homologie der Keimblätter. Jena. Z. Naturw. **8**,1–55.

HARBISON, G. R. (1985). On the classification and evolution of the Ctenophora. In S. CONWAY MORRIS, J. D. GEORGE, R. GIBSON & H. M. PLATT (Eds.). The origins and relationships of lower invertebrates. The Systematics Ass. Spec. Vol. **28**, 78–100. Clarendon Press. Oxford.

HARBISON, G. R. & R. S. MILLER (1986). Not all ctenophores are hermaphrodites. Studies on the systematics, distribution, sexuality and development of two species of Ocyropsis. Mar. Biol. **90**, 413–424.

HARRISON, F. W. & L. DE VOS (1991). Porifera. In F. W. HARRISON & J. A. WESTFALL (Eds.). Microscopic Anatomy of Invertebrates **2**, 29–89. Wiley-Liss, Inc. New York.

HASZPRUNAR, G., RIEGER, R. & P. SCHUCHERT (1991). On the origin of the Bilateria: traditional views and recent alternative concepts. In A. M. SIMONETTA & S. CONWAY MORRIS (Eds.). The early evolution of Metazoa and the significance of problematic taxa. 107–112. Cambridge University Press. Cambridge.

HATSCHEK, B. (1878). Studien über die Entwicklungsgeschichte der Annelida. Arbeiten Zool. Inst. Wien **1**, 277–404.

HENDELBERG, J. & K.-O. HEDLUND (1974). On the morphology of the epidermal ciliary rootlet system of the acoelous turbellarian Childia groenlandica. Zoon **2**, 13–24.

HENNIG, W. (1950). Grundzüge einer Theorie der phylogenetischen Systematik. Deutscher Zentralverlag. Berlin.

HENNIG, W. (1966). Phylogenetic Systematics. University of Illinois Press. Urbana, Chicago, London.

HENNIG, W. (1994). Wirbellose I. Taschenbuch der Speziellen Zoologie. Teil 1. UTB 1831. G. Fischer. Jena.

HENTSCHEL, E. & G. WAGNER (1990). Zoologisches Wörterbuch. UTB 367. G. Fischer. Stuttgart.

HERNANDEZ-NICAISE, M.-L. (1991). Ctenophora. In F. W. HARRISON & J. A. WESTFALL (Eds.). Microscopic Anatomy of Invertebrates **2**, 359–418. Wiley-Liss, Inc. New York.

HIBBERD, D. J. (1975). Observations on the ultrastructure of the choanoflagellate Codosiga botrytis (Ehr.) Saville-Kent with a special reference to the flagellar apparatus. J. Cell. Sci. **17**, 191–219.

HILLIS, D. M. (1987). Molecular versus morphological approaches to Systematics. Annu. Rev. Ecol. Syst. **18**, 23–42.

HILLIS, D. M. & C. MORITZ (Eds.) (1990). Molecular Systematics. Sinauer Associates Inc. Sunderland, Massachusetts.

HOFFMANN, H. (1930). Amphineura und Scaphopoda. Nachträge. Bronns Klass. u. Ordn. d. Tierreichs **3**, I, 511p. Akad. Verlagsges. Leipzig.

HOLSTEIN, T. & K. HAUSMANN (1988). The cnidocil apparatus of hydrozoans: A progenitor of higher metazoans mechanoreceptors? In D. A. HESSINGER and H. M. LENHOFF (Eds.). The Biology of Nematocysts. 53–73. Academic Press, Inc. San Diego.

HULL, D. L. (1976). Are species really individuals? Syst. Zool. **25**, 174–191.

HYMAN, L. H. (1940). The Invertebrates: Protozoa through Ctenophora. McGraw-Hill Book Comp. New York, London.

HYMAN, L. H. (1951). The Invertebrates: Platyhelminthes and Rhynchocoela. The acoelomate Bilateria. McGraw-Hill Book Comp. New York, Toronto, London.

JÄGERSTEN, G. (1972). Evolution of the metazoan life cycle. Academic Press. London, New York.

JEFFERIES, R. P. S. (1986). The ancestry of the Vertebrates. British Museum (Natural History). London.

JESPERSEN, A. & J. LÜTZEN (1987). Ultrastructure of the nephridiocirculatory connections in Tubulanus annulatus (Nemertini, Anopla). Zoomorphology **107**, 181–189.

JOYEUX, C. & J. G. BAER (1961). Classe des Cestodes. Traité de Zoologie **4** (1), 347–560.

KAESTNER, A. (1993). Lehrbuch der Speziellen Zoologie. E. GRUNER (Hrg.). Wirbellose Tiere. 1. Teil: Einführung, Protozoa, Placozoa, Porifera. G. Fischer. Jena, Stuttgart, New York.

KÄMPFE, L. (Hrg.) (1992). Evolution und Stammesgeschichte der Organismen. G. Fischer. Stuttgart.

KARLING, T. G. (1940). Zur Morphologie und Systematik der Alloeocoela Cumulata und Rhabdocoela Lecithophora (Turbellaria). Acta Zool. Fenn. **26**, 1–260.

KARLING, T. G. (1974). On the anatomy and affinities of the turbellarian orders. In N. W. Riser & M. P. Morse (Eds.). Biology of the Turbellaria. 1–16. McGraw-Hill Book Company. New York.

KLEINIG, H. & P. SITTE (1992). Zellbiologie. G. Fischer. Stuttgart, Jena, New York.

KORN, H. (1960). Zur Dauer der Metamorphose von Echiurus abyssalis Skor. (Echiurida, Annelida). Kieler Meeresforsch. **16**, 238–242.

KORSCHELT, E. & K. HEIDER (1892). Lehrbuch der vergleichenden Entwicklungsgeschichte der wirbellosen Tiere; spezieller Teil. G. Fischer. Jena.

KORSCHELT, E. & K. HEIDER (1895). Text-book of the embryology of Invertebrates. Part I. Swan Sonnenschein. New York, London.

KOZLOFF, E. N. (1965). Ciliocincta sabellariae gen. and sp. n., an orthonectid mesozoan from the polychaete Sabellaria cementarium Moore. J. Parasit. **51**, 37–44.

KOZLOFF, E. N. (1969). Morphology of the orthonectid Rhopalura ophiocomae. J. Parasit. **55**, 171–195.

KOZLOFF, E. N. (1971). Morphology of the orthonectid Ciliocincta sabellariae. J. Parasit. **57**, 585–597.

KOZLOFF, E. N. (1990). Invertebrates. Saunders College Publishing. Philadelphia.

KOZLOFF, E. N. (1992). The genera of the phylum Orthonectida. Cah. Biol. Mar. **33**, 377–406.

KÜMMEL, G. (1977). Der gegenwärtige Stand der Forschung zur Funktionsmorphologie exkretorischer Systeme. Versuch einer vergleichenden Darstellung. Verh. Deutsch. Zool. Ges. 1977, 154–174.

LAFAY, B., BOURY-ESNAULT, N., VACELET, J. & B. CHRISTEN (1992). An analysis of partial 28S ribosomal RNA sequences suggests early radiations of sponges. BioSystems **28**, 139–151.

LAKE, J. A. (1990). Origin of the Metazoa. Proc. Natl. Acad. Sci. USA **87**, 763–766.

LAMMERT, V. (1984). The fine structure of spiral ciliary receptors in Gnathostomulida. Zoomorphology **104**, 360–364.

LAMMERT, V. (1986). Vergleichende Ultrastruktur-Untersuchungen an Gnathostomuliden und die phylogenetische Bewertung ihrer Merkmale. Dissertation Univ. Göttingen.

LAMMERT, V. (1989). Fine structure of the epidermis in Gnathostomulida. Zoomorphology **109**, 131–144.

LAMMERT, V. (1991). Gnathostomulida. In F. W. HARRISON & E. E. RUPPERT. (Eds.). Microscopic Anatomy of Invertebrates **4**, 19–39. Wiley-Liss, Inc. New York.

LANG, A. (1884). Die Polycladen (Seeplanarien) des Golfes von Neapel und der angrenzenden Meeresabschnitte. Fauna und Flora des Golfes von Neapel **11**, 668pp. W. Engelmann. Leipzig.

LANG, A. (1891). Text-book of comparative anatomy. Part 1. Macmillan. London, New York.

LAPAN, E. A. & H. MOROWITZ (1972). The Mesozoa. Scientific American **277**, 94–101.

LASKA-MEHNERT, G. (1985). Cytologische Veränderungen während der Metamorphose des Cubopolypen Tripedalia cystophora (Cubozoa, Carybdeidae) in die Meduse. Helgoländer Meeresunters. **39**, 129–164.

LEEDALE, G. F. (1974). How many are the kingdoms of organisms? Taxon **23**, 261–270.

LEIPE, D. & K. HAUSMANN (1993). Neue Erkenntnisse zur Stammesgeschichte der Eucaryoten. Biologie in unserer Zeit **23**, 178–183.

LEVIN, N. D., CORLISS, J. O., COX, F. E. G., DEROUX, G., GRAIN, J., HONIGBERG, B. M., LEEDALE, G. F., LOEBLICH, A. R., LOM III, J., LINN, D., MERINFELD, E. G., PAGE, F. C., POLALJANSKY, G., SPRAGUE, V., VAVRA, J. & F. G. WALLACE (1980). A newly revised classification of the Protozoa. J. Protozool. **27**, 37–58.

LINNAEUS, C. (1758). Systema naturae per regna tria naturae, secundum classes, ordines, genera, species cum caracteribus, differentiis, synonymis, locis. Editio decima reformata. Tom I. Laurentii Salvii, Holmae.

LUTHER, A. (1905). Zur Kenntnis der Gattung Macrostoma. Festschrift für Palmén. No. **5**, 1–61. Helsingfors.

LUTHER, A. (1960). Die Turbellarien Ostfennoskandiens. I. Acoela, Catenulida, Macrostomida, Lecithoepitheliata, Prolecithophora und Proseriata. Fauna Fennica **7**, 1–155.

MAHNER, M. (1993). What is a species? A contribution to the never ending species debate in biology. Journal for General Philosophy of Science **24**, 103–126.

MAHNER, M. (1994). Phänomenalistische Erblast in der Biologie. Biol. Zent. bl. **113**, 435–448.

MAHNER, M. & M. BUNGE. Philosophy of Biology. Manuscript.

MARGULIS, L. (1981). Symbiosis in cell evolution. W. H. Freemann and Company. San Francisco.

MARGULIS, L. & K. V. SCHWARTZ (1988). Five kingdoms. An illustrated guide to the phyla of life on earth. W. H. Freemann. New York.

MARGULIS, L. & K. V. SCHWARTZ (1989). Die fünf Reiche der Organismen. Ein Leitfaden. Spektrum der Wissenschaft. Verlagsgesellschaft mbH & Co. Heidelberg.

MATSUBARA, J. A. & P. L. DUDLEY (1976a). Fine structural studies of the dicyemid mesozoan, Dicyemmenea californica McConnaughey. I. Adult stages. J. Parasit. **62**, 377–389.

MATSUBARA, J. A. & P. L. DUDLEY (1976b). Fine structural studies of the dicyemid mesozoan, Dicyemmenea californica McConnaughey. II. The young vermiform stage and the infusoriform larvae. J. Parasit. **62**, 390–409.

MAYR, E. (1975). Grundlagen der zoologischen Systematik. Parey. Hamburg, Berlin.

MAYR, E. (1982). The growth of biological thought. Diversity, evolution, and inheritance. Harvard University Press. Cambridge, Massachusetts.

McCONNAUGHEY, B. H. (1951). The life cycle of the dicyemid Mesozoa. Univ. Californ. Publ. in Zoology **55**, 295–335.

MEGLITSCH, P. A. & F. R. SCHRAM (1991). Invertebrate Zoology. Oxford University Press. New York, Oxford.

MEIER, R. (1992). Der Einsatz von Computern in phylogenetischen Analysen – eine Übersicht. Zool. Anz. **229**, 106–133.

MEHL, D. & H. M. REISWIG (1991). The presence of flagellar vanes in choanomers of Porifera and their possible phylogenetic implications. Z. zool. Syst. Evolut.-forsch. **29**, 312–319.

MEHL, D. & J. REITNER (1991). Monophylie und Systematik der Porifera. Verh. Deutsch. Zool. Ges. **84**, 447.

MEYER, A. (1926). Logik der Morphologie. J. Springer, Berlin.

MEYER, A. (1993). Molecular approaches to the phylogenetic study of vertebrates. Verh. Deutsch. Zool. Ges. **86**. 2, 131–149.

MÖHN, E. (1984). System und Phylogenie der Lebewesen 1. Schweizerbart'sche Verlagsbuchhandlung. Stuttgart.

MOORE, J. & R. GIBSON (1993). Methods of classifying nemerteans: an assessment. Hydrobiologia **266**, 89–101.

MÜLLER, U. & P. AX (1971). Gnathostomulida von der Nordseeinsel Sylt mit Beobachtungen zur Lebensweise und Entwicklung von Gnathostomula paradoxa Ax. Mikrofauna Meeresboden **9**, 1–41.

MÜLLER, W. E. G., MÜLLER, I. M., RINKEVICH, B. & V. GAMULIN (1995). Molecular Evolution: Evidence for the monophyletic origin of multicellular animals. Naturwissenschaften **82**, 36–38.

NAWITZKI, W. (1931). Procarinina remanei. Eine neue Paläonemertine der Kieler Förde. Zool. Jahrb. Anat. **54**, 159–234.

NEUHAUS, B. (1994). Ultrastructure of alimentary canal and body cavity, ground pattern and phylogenetic relationships of the Kinorhyncha. Microfauna Marina **9**, 61–156.

NORENBURG, J. L. (1988). Remarks on marine interstital nemertines and key to the species. Hydrobiologia **156**, 87–92.

PALMBERG, I. (1990). Stem cells in microturbellarians. An autoradiographic and immunocytochemical study. Protoplasma **158**, 109–120.

PATTERSON, C. & D. E. ROSEN (1977). Review of ichthyodectiform and other mesozoic teleost fishes and the theory and practice of classifying fossils. Bull. Am. Mus. Natur. Hist. **158**, 81–172.

PATTERSON, C., WILLIAMS, D. M. & C. J. HUMPHRIES (1993). Congruences between molecular and morphological phylogenies. Annu. Rev. Ecol. Syst. **24**, 153–188.

PEDERSEN, K. J. & L. R. PEDERSEN (1986). Fine structural observations on the extracellular matrix (ECM) of Xenoturbella bocki Westblad, 1949. Acta Zool. (Stockh.) **67**, 103–113.

PEDERSEN, K. J. & L. R. PEDERSEN (1988). Ultrastructural observations on the epidermis of Xenoturbella bocki Westblad, 1949, with a discusion of epidermal cytoplasmic filament systems of Invertebrates. Acta Zool. (Stockh.) **69**, 231–246.

PETERSEN, K. W. (1979). Development of coloniality in Hydrozoa. In G. LAERWOOD & B. R. ROSEN (Eds.). Biology and systematics of colonial organisms. Systematics Association Spec. Vol. **11**, 105–139. Academic Press. London.

PETERSEN, K. W. (1990). Evolution and taxonomy in capitate hydroids and medusae (Cnidaria: Hydrozoa). Zool. J. Linn. Soc. **100**, 101–231.

PIEKARSKI, G. (1954). Lehrbuch der Parasitologie. Springer. Berlin, Göttingen, Heidelberg.

PITELKA, D. R. (1974). Basal bodies and root structures. In M. A. SLEIGH (Ed.). Cilia and Flagella. 437–469. Academic Press. London, New York.

RAFF, R. A., MARSHALL, C. R. & J. M. TURBEVILLE (1994). Using DNA sequences to unravel the cambrian radiation of the animal phyla. Annu. Rev. Ecol. Syst. **25**, 351–375.

REISINGER, E. (1924). Die Gattung Rhynchoscolex. Z. Morph. Ökol. Tiere **1**, 1–37.

REMANE, A. (1952). Die Grundlagen des natürlichen Systems, der vergleichenden Anatomie und der Phylogenetik. Akademische Verlagsgesellschaft Geest & Portig. Leipzig.

RIDLEY, R. K. (1968). Electron microscopic studies on dicyemid Mesozoa. I. Vermiform stages. J. Parasit. **54**, 975–998.

RIDLEY, R. K. (1969). Electron microscopic studies on dicyemid Mesozoa. II. Infusorigen and Infusoriform stages. J. Parasit. **55**, 779–793.

RIEDL, R. (1969). Gnathostomulida from America. Science **163**, 445–452.

RIEGER, R. M. (1976). Monociliated epidermal cells in Gastrotricha. Significance for concepts of early metazoan evolution. Z. zool. Syst. Evolut.-forsch. **14**, 198–226.

RIEGER, R. (1986). Über den Ursprung der Bilateria: die Bedeutung der Ultrastrukturforschung für ein neues Verstehen der Metazoenevolution. Verh. Deutsch. Zool. Ges. **79**, 31–50.

RIEGER, R. M. (1994). Evolution of the „lower" Metazoa. In S. BENGTSON (Ed.). Early life on Earth. Nobel Symp. **84**, 475–488. Columbia University Press. New York.

RIEGER, R. M. & J. LOMBARDI (1987). Ultrastructure of coelomic lining in echinoderm podia: significance for concepts in the evolution of muscle and peritoneal cells. Zoomorphology **107**, 191–208.

RIEGER, R. M. & M. MAINITZ (1977). Comparative fine structural study of the body wall in Gnathostomulids and their phylogenetic position between Platyhelminthes and Aschelminthes. Z. zool. Syst. Evolut.-forsch. **15**, 9–35.

RISER, N. W. (1988). Arhynchonemertes axi gen. n., sp. n. (Nemertinea) – an insight into basic acoelomate bilaterian organology. In P. AX, U. EHLERS & B. SOPOTT-EHLERS (Eds.). Free-living and symbiotic Plathelminthes. Fortschritte der Zoologie/Progress in Zoology **36**, 367–373.

RIUTORT, M., FIELD, K. G., RAFF, R. A. & J. BAGUÑÀ (1993). 18S rRNA and phylogeny of Plathelminthes. Biochem. Syst. Ecol. **21**, 71–77.

RÖD, W. (1988). Die Philosophie der Antike 1: Von Thales bis Demokrit. Ch. Beck, München.

ROHDE, K., WATSON, N. & L. R. G. CANNON (1988). Ultrastructure of epidermal cilia of Pseudactinoposthia sp. (Platyhelminthes, Acoela); implications for the phylogenetic status of the Xenoturbellida and Acoelomorpha. J. Submicros. Cytol. Pathol. **20**, 759–767.

RUPPERT, E. E. (1978). A review of metamorphosis of turbellarian larvae. In CHIA & M. RICE (Eds.). Settlement and metamorphosis of marine invertebrate larvae. 65–81. Elsevier. Amsterdam.

RUPPERT, E. E. (1991). Introduction to the Aschelminth phyla: A consideration of mesoderm, body cavities and cuticle. In F. W. HARRISON & E. E. RUPPERT (Eds.). Microscopic Anatomy of Invertebrates **4**, 1–17. Wiley-Liss, Inc. New York.

RUTHMANN, A., BEHRENDT, G. & R. WAHL (1986). The ventral epithelium of Trichoplax adhaerens (Placozoa): Cytosceletal structures, cell contacts and endocytosis. Zoomorphology **106**, 115–122.

SALVINI-PLAWEN, L. v. (1978). On the origin and evolution of the lower Metazoa. Z. zool. Syst. Evolut.-forsch. **16**, 40–88.

SALVINI-PLAWEN, L. v. (1980). Was ist eine Trochophora? Eine Analyse der Larventypen mariner Protostomier. Zool. Jb. Anat. **103**, 389–423.

SCHMIDT, H. (1972). Die Nesselkapseln der Anthozoen und ihre Bedeutung für die phylogenetische Systematik. Helgoländer wiss. Meeresunters. **23**, 422–458.

SCHMIDT, H. (1974). On evolution in the Anthozoa. Proc. Second. Int. Coral Reef. Symp. **1**, 533–560. Great Barrier Reef Committee. Brisbane.

SCHMIDT, H. & B. MORAW (1982). Die Cnidogenese der Octocorallia (Anthozoa, Cnidaria): II. Reifung, Wanderung und Zerfall von Cnidoblast und Nesselkapsel. Helgoländer Meeresunters. **35**, 97–118.

SCHMIDT-RHAESA, A. (1993). Ultrastructure and development of the spermatozoa of Multipeniata (Plathelminthes, Prolecithophora). Microfauna Marina **8**, 131–138.

SCHUCHERT, P. (1993). Phylogenetic analysis of the Cnidaria. Z. zool. Syst. Evol.-forsch. **31**, 161–173.

SCHWEMMLER, W. (1979). Mechanismen der Zellevolution. W. de Gruyter. Berlin, New York.

SIEWING, R. (1969). Lehrbuch der vergleichenden Entwicklungsgeschichte der Tiere. Parey. Hamburg, Berlin.

SIEWING, R. (1985). In H. WURMBACH & R. SIEWING. Lehrbuch der Zoologie. Band 2. Systematik. G. Fischer. Stuttgart, New York.

SMITH, A. B. (1992). Echinoderm Phylogeny: Morphology and molecules approach accord. Trends Ecol. Evol. **7**, 224–229.

SMITH, P. R. & E. E. RUPPERT (1988). Nephridia. In W. WESTHEIDE & C. O. HERMANS (Eds.). The Ultrastructure of Polychaeta. Microfauna Marina **4**, 231–262. G. Fischer. Stuttgart, New York.

SMITH III, J. & S. TYLER (1985). The acoel turbellarians: kingpins of metazoan evolution or a specialized offshoot? In S. CONWAY MORRIS, J. B. GEORGE, R. GIBSON & H. M. PLATT (Eds.). The origins and relationships of lower invertebrates. The Systematics Ass. Special Vol. **28**, 123–142. Clarendon Press. Oxford.

SOGIN, M. L. (1994). The origin of eucaryotes and evolution into major kingdoms. In S. BENGTSON (Ed.). Early life on Earth. Nobel Symp. **84**, 181–192. Columbia University Press. New York.

SOPOTT-EHLERS, B. (1985). The phylogenetic relationships within the Seriata (Plathelminthes). In S. CONWAY MORRIS, J. D. GEORGE, R. GIBSON & H. M. PLATT (Eds.). The origins and relationships of lower invertebrates. The Systematics. Ass. Special Vol. **28**, 159–167. Clarendon Press. Oxford.

STERRER, W. (1968). Beiträge zur Kenntnis der Gnathostomulida. I. Anatomie und Morphologie des Genus Pterognathia Sterrer. Ark. f. Zool. Ser. 2. **22**, 1–125.

STERRER, W. (1972). Systematics and evolution within the Gnathostomulida. Syst. Zool. **21**, 151–173.

STERRER, W. (1991a). Gnathostomulida from Fiji, Tonga and New Zealand. Zool. Scripta **20**, 107–128.

STERRER, W. (1991b). Gnathostomulida from Hawaii. Zool. Scripta **20**, 129–136.

STERRER, W. (1991c). Gnathostomulida from Tahiti. Zool. Scripta **20**, 137–146.

STERRER, W. & R. RIEGER (1974). Retronectidae – a new cosmopolitan marine family of Catenulida (Turbellaria). In N. W. RISER & M. P. MORSE (Eds.). Biology of the Turbellaria. 63–92. McGraw-Hill Book Company. New York.

STORCH, V. & U. WELSCH (1991). Systematische Zoologie. G. Fischer. Stuttgart, New York.

STORCH, V. & U. WELSCH (1993). Kükenthals Leitfaden für das Zoologische Praktikum. G. Fischer. Stuttgart, Jena.

STRICKER, S. A. & R. A. CLONEY (1982). Stylet formation in Nemerteans. Biol. Bull. **162**, 387–403.

STUNKARD, H. W. (1954). The life-history and systematic relations of the Mesozoa. Quart. Rev. Biol. **29**, 230–244.

SUDHAUS, W. & K. REHFELD (1992). Einführung in die Phylogenetik und Systematik. G. Fischer, Stuttgart.

SUNDBERG, P. (1993). Phylogeny, natural groups and nemertean classification. Hydrobiologia **266**, 103–113.

TARDENT, P. (1978). Coelenterata, Cnidaria. In F. Seidel (Ed.). Morphogenese der Tiere. Erste Reihe. Lieferung 1: A–I, 69–415. G. Fischer. Stuttgart, New York.

THIEL, H. (1966). The evolution of Scyphozoa. A review. In W. J. REES (Ed.). The Cnidaria and their evolution. 77–117. Academic Press. London.

THOMAS, M. B. (1975). The structure of the 9+1 axonemal core as revealed by treatment with trypsin. J. Ultrastr. Res. **52**, 409–422.

THOMAS, M. B. & N. C. EDWARDS (1991). Cnidaria: Hydrozoa. In F. W. HARRISON & J. A. WESTFALL (Eds.). Microscopic Anatomy of Invertebrates **2**, 91–183. Wiley-Liss, Inc. New York.

TURBEVILLE, J. M. (1991). Nemertinea. In F. W. HARRISON & B. J. BOGITSH (Eds.). Microscopic Anatomy of Invertebrates **3**, 285–328. Wiley-Liss, Inc. New York.

TURBEVILLE, J. M., FIELD, K. G. & R. R. RAFF (1992). Phylogenetic position of phylum Nemertini, inferred from 18S rRNA sequences: Molecular data as a test of morphological character homology. Mol. Biol. Evol. **9**, 235–249.

TURBEVILLE, J. M. & E. E. RUPPERT (1985). Comparative ultrastructure and evolution of Nemertines. Amer. Zool. **25**, 53–71.

TYLER, S. (1976). Comparative ultrastructure of adhesive systems in the Turbellaria. Zoomorphologie **84**, 1–76.

VAN DE PEER, Y., NEEFS, J.-M., DE RIJK, P. & R. DE WACHTER (1993). Evolution of Eucaryotes as deduced from small ribosomal subunit RNA sequences. Biochem. Syst. Ecol. **21**, 43–55.

VERGER-BOCQUET, M. (1992). Polychaeta: Sensory Structures. In F. W. HARRISON & S. L. GARDINER (Eds.). Microscopic Anatomy of Invertebrates **7**, 181–196. Wiley-Liss, Inc. New York.

VICKERMAN, K., BRUGEROLLE, G. & J.-P. MIGNOT (1991). Mastigophora. In F. W. HARRISON & J. O. CORLISS (Eds.). Microscopic Anatomy of Invertebrates **1**, 13–159. Wiley-Liss, Inc. New York.

VOLLMER, G. (1975). Evolutionäre Erkenntnistheorie. S. Hirzel, Stuttgart.

WÄGELE, J. W. (1994a). Rekonstruktion der Phylogenese mit DNA-Sequenzen: Anspruch und Wirklichkeit. Natur und Museum **124**, 225–231.

WÄGELE, J. W. (1994b). Review of methological problems of „Computer cladistics" exemplified with a case study on isopod phylogeny (Crustacea: Isopoda). Z. zool. Syst. Evolut.-forsch. **32**, 82–107.

WÄGELE, J. W. & R. WETZEL (1994). Nucleic acid sequence data are not per se reliable for inference of phylogenies. J. Nat. Hist. **28**, 749–761.

WAINRIGHT, P. O., HINKE, G., SOGIN, M. L. & S. K. STICKEL (1993). Monophyletic origin of the Metazoa: An evolutionary link with Fungi. Science **260**, 340–342.

WEHNER, R. & W. GEHRING (1990). Zoologie. G. Thieme. Stuttgart, New York.

WEISSENFELS, N. (1989). Biologie und Mikroskopische Anatomie der Süßwasserschwämme (Spongillidae). G. Fischer. Stuttgart, New York.

WEISSENFELS, N. (1992). The filtration apparatus for food collection in freshwater sponges (Porifera, Spongillidae). Zoomorphology **112**, 51–55.

WERNER, B. (1993). Stamm Cnidaria. In H.-E. GRUNER (Ed.). Lehrbuch der Speziellen Zoologie. Band 1: Wirbellose Tiere. 2. Teil, 11–305. G. Fischer. Jena, Stuttgart, New York.

WESTBLAD, E. (1937). Die Turbellarien-Gattung Nemertoderma Steinböck. Acta Soc. Fauna et Flora Fenn. **60**, 45–89.

WESTBLAD, E. (1949). Xenoturbella bocki n. g., n. sp., a peculiar, primitive Turbellarian type. Ark. f. Zool. **1**, 11–29.

WHITTAKER, E. H. (1959). On the broad classification of organisms. Quart. Rev. Biol. **34**, 210–226.

WHITTAKER, E. H. (1969). New concepts of the kingdoms of organisms. Science **163**, 150–160.

WILLMANN, R. (1985). Die Art in Raum und Zeit. P. Parey. Berlin, Hamburg.

WOLFE, J. (1972). Basal body fine structure and chemistry. Adv. Cell. molec. Biol. **2**, 151–192.

WURMBACH, H. & R. SIEWING (1985). Lehrbuch der Zoologie. Band 2. Systematik. G. Fischer. Stuttgart, New York.

Register

(Names of non-monophyletic taxa in quotation marks)

222

Comparing structure with function

Zoomorphology
An International Journal of
Comparative and Functional
Morphology
ISSN 0720-213X
Title No. 435

Managing Editor:
O. Kraus

In cooperation with an
international Editorial Board

The journal covers research on morphological investigations of invertebrates and vertebrates at the macroscopic, microscopic and ultrastructural levels, including embryological studies.

Special emphasis is placed on:

- Comparative anatomical studies that correlate structure with function, including morphometric analysis;

- Analysis of interrelationships between structural-functional systems of animals and their general biology, including environmental adaptations and behavior;

- Analysis of interdependency among complex structural-functional systems in adult organisms, as well as during embryological and phylogenetical development;

- Studies of developmental phenomena and homologies as the basis for phylogenetic relationships.

Subscription information for 1996:

Volume 116 (4 issues)
DM 1125,–*

* plus
carriage
charges.
In EU countries
the local VAT
is effective.

Springer

Springer-Verlag, P. O. Box 31 13 40, D-10643 Berlin, Germany.

IMCA.3304/MNTZ/SF